THE LIBRARY
ST. MARY'S COLLEGE OF MARYLAND
ST. MARY'S CITY, MARYLAND 20686

083073

Regulation of Parasite Populations

ACADEMIC PRESS RAPID MANUSCRIPT REPRODUCTION

Regulation of Parasite Populations

EDITED BY

Gerald W. Esch

*Department of Biology
Wake Forest University
Winston-Salem, North Carolina*

with
Introductory Remarks by
BRENT B. NICKOL
*School of Life Sciences
University of Nebraska-Lincoln
Lincoln, Nebraska*

ACADEMIC PRESS, INC.   New York   London   San Francisco   1977
A Subsidiary of Harcourt Brace Jovanovich, Publishers

COPYRIGHT © 1977, BY ACADEMIC PRESS, INC.
ALL RIGHTS RESERVED.
NO PART OF THIS PUBLICATION MAY BE REPRODUCED OR
TRANSMITTED IN ANY FORM OR BY ANY MEANS, ELECTRONIC
OR MECHANICAL, INCLUDING PHOTOCOPY, RECORDING, OR ANY
INFORMATION STORAGE AND RETRIEVAL SYSTEM, WITHOUT
PERMISSION IN WRITING FROM THE PUBLISHER.

ACADEMIC PRESS, INC.
111 Fifth Avenue, New York, New York 10003

United Kingdom Edition published by
ACADEMIC PRESS, INC. (LONDON) LTD.
24/28 Oval Road, London NW1

Library of Congress Cataloging in Publication Data

Main entry under title:

Regulation of parasite populations.

    Proceedings of a symposium held at New Orleans, November 10-14, 1975, and jointly sponsored by the American Microscopial Society and the American Society of Parasitologists.
    Includes index.
    1. Parasitology—Congresses.   2. Parasites—Control—Congresses.   I. Esch, Gerald W.   II. American Microscopial Society.   III. American Society of Parasitologists.   [DNLM: 1. Parasites—Congresses. 2. Population dynamics—Congresses.   3. Models, Biological—Congresses.   4. Ecology—Congresses. 5. Pest control—Congresses.   QX4 R344 1975]
QH547.R43   574.5'24   76-53012
ISBN 0–12–241750–X

PRINTED IN THE UNITED STATES OF AMERICA

*Because of our respect for them,
this volume is dedicated to:
Professor J. Teague Self
and
Professor Emeritus Robert M. Stabler*

Contents

List of Contributors	*ix*
Preface	*xi*
INTRODUCTORY REMARKS Brent B. Nickol	1
PARASITISM AND r- and K-SELECTION Gerald W. Esch, Terry C. Hazen, and John M. Aho	9
THE REGULATION OF FISH PARASITE POPULATIONS C. R. Kennedy	63
THE ROLE OF ARRESTED DEVELOPMENT IN THE REGULATION OF NEMATODE POPULATIONS Gerhard A. Schad	111
USE OF MATHEMATICAL MODELS IN PARASITOLOGY Robert P. Hirsch	169
POPULATIONS IN PERSPECTIVE: COMMUNITY ORGANIZATION AND REGULATION OF PARASITE POPULATIONS John C. Holmes, Russell P. Hobbs, and Tak Seng Leong	209
Index	*247*

List of Contributors

JOHN M. AHO, Department of Biology, Wake Forest University, Winston-Salem, North Carolina, U. S. A. (27109)

GERALD W. ESCH, Department of Biology, Wake Forest University, Winston-Salem, North Carolina, U. S. A. (27109)

TERRY C. HAZEN, Department of Biology, Wake Forest University, Winston-Salem, North Carolina, U. S. A. (27109)

ROBERT P. HIRSCH, Division of Biology, Kansas State University, Manhattan, Kansas, U. S. A. (66506)

RUSSELL P. HOBBS, Department of Zoology, University of Alberta, Edmonton, Alberta, Canada (T6G 2E9)

JOHN C. HOLMES, Department of Zoology, University of Alberta, Edmonton, Alberta, Canada (T6G 2E9)

C. R. KENNEDY, Department of Biological Sciences, Hatherly Laboratories, Prince of Wales Road, Exeter, United Kingdom (EX4 4PS)

TAK SENG LEONG, Department of Zoology, University of Alberta, Edmonton, Alberta, Canada (T6G 2E9)

BRENT B. NICKOL, School of Life Sciences, University of Nebraska-Lincoln, Lincoln, Nebraska, U. S. A. (68508)

GERHARD A. SCHAD, School of Veterinary Medicine, University of Pennsylvania, 380 Spruce Street H-1, Philadelphia, Pennsylvania, U. S. A. (19174)

Preface

In November, 1975, a symposium was jointly sponsored by the American Microscopical Society and the American Society of Parasitologists at their meeting in New Orleans, Louisiana. The purpose of the symposium was to focus on past and current literature dealing with regulation of parasite populations as well as to introduce some new concepts and notions regarding this particular area of interest. The outcome of this effort was the blending of new and old ideas held by workers in both ecology and parasitology. While some of the approaches taken by the participants will be recognized as dogma by many workers, in either ecology or parasitology, much of the same information will be new, and hopefully even provocative, to many others in both areas. In a sense, therefore, the symposium and this volume represent an effort to "bridge a gap" between some of the current ideas and thinking in ecology and parasitology. Synthesis is never easy to accomplish, but for many in both disciplines, the symposium contributions presented herein should be both illuminating and stimulating.

As is the case for any volume of this kind, a number of people made significant contributions of one sort or another. These include, Drs. L. Roberts and G. Schmidt, Program Officers for the American Microscopical Society and American Society of Parasitologists, respectively. We also appreciate the help of Mr. Robert Nestor and Dr. Michael Smith of the Savannah River Ecology Laboratory for their cooperation in assisting with the final preparation of the volume. We also want especially to thank Ms. Tonya Willingham and Mrs. Gloria Weiner for their most generous help in typing the final drafts of all six manuscripts. Mr. Terry C. Hazen and Mr. John Aho assisted with proofing the final copy.

In part, the preparation of the volume was supported by Contract E(38-1)-900 between Wake Forest University and the Energy Research and Development Administration and Contract E(38-1)-819 between the University of Georgia and the Energy Research and Development Administration.

Introductory Remarks

BRENT B. NICKOL

School of Life Sciences
University of Nebraska-Lincoln
Lincoln, Nebraska
U.S.A.

When we first begin to talk about parasites and parasitism to freshman zoology classes or in beginning parasitology courses, hazards in the parasitic mode of life are frequently emphasized. We clearly recognize risks in transmission from host to host encountered by parasites and dwell on vulnerability of free-living stages to the environment, elaborate mechanisms of host location, and defensive responses by the host. All these pitfalls to successful completion of a reproductive cycle are then gloriously reconciled by pointing out the high reproductive potential of most parasites. According to Crompton and Whitfield (1968), female *Polymorphus minutus* each produce 1700 eggs per day, *Moniliformis moniliformis* about 5500 (Crompton, Arnold, and Barnard, 1972), and Kates (1944) tells us that each female *Macracanthorhynchus hirudinaceus* produces more than a quarter of a million eggs each day. These totals are equalled or exceeded by many species of nematodes; polyembryony occurs among digenetic trematodes; and cestodes lead a life of perpetual rejuvenation. Indeed, one might wonder why parasites do not continually increase in number.

There are obviously upper and lower limits between which population densities for any species must remain if the species is to survive. Den Boer (1968) believed

that limitation of density has little influence on the
magnitude of density fluctuations and felt that population regulation is not so much a problem of limiting
density as it is one of restricting density fluctuations. He contended that by "spreading the risk",
density fluctuations could be leveled out without the
mandatory need for density-dependent regulatory mechanisms. Phenotypic variation, variation in time, variation in space, and variation in relations with other
species are means by which fluctuations in density are
stabilized.

Stabilization by phenotypic variation is exemplified by a species of carabid beetle in which the number of pits on the elytra is correlated with sensitivity of larvae for the moisture content of the substrate
(Den Boer, 1968). Spreading the risk through time results from individual variation in time and rate of
development and/or reproduction. This mechanism may
function through the varied rate of development displayed by several parasite species when heavy infection
of intermediate hosts occurs. DeGiusti (1949) reported
that heavy infections of *Leptorhynchoides thecatus* in
Hyalella azteca result in larvae of various stages of
development. Similar results are known for other helminth species including *Prosthorhynchus formosus* in
Armadillidium vulgarae in our laboratory. This means
of spreading the risk by variation in time could be
especially effective if infectivity of fully formed
stages were of limited duration. Although it seems
likely that these particular examples are density-dependent, one might visualize a similar density-independent mechanism. Arrested development of larval nematodes, a topic explored later in this symposium, could
result in stabilization of this sort.

Stabilization resulting from variation in space is
purportedly brought about because effective environments of natural populations are heterogeneous and extreme conditions in one place balance, to some degree,
less extreme conditions in different microenvironments.
Perhaps among parasites this could be extended to include utilization of multiple definitive hosts. Later
in this volume, however, Holmes demonstrates that parasite density fluctuations in a principal definitive
host are not leveled by less extreme conditions in

other hosts. Finally, according to Den Boer, a reduction in the amplitude of fluctuation of animal numbers in the populations concerned can result from heterogeneous relations with other species such as in the case of polyphagous predators.

Den Boer (1968) does not contend that regulation by his "spreading the risk" excludes the possibility of additional stabilization by some density-dependent regulatory mechanism. Irregular fluctuation of most wild animals, including parasites, in numbers between limits that are extremely restricted compared with what their rates of increase would allow, implies regulatory mechanisms. Regulation by a combination of high fecundity and high mortality, as alluded to for parasites, is expensive to operate and is a relatively poor regulator because it tends to produce strong fluctuations. Yet as Bradley (1974) has pointed out, host-parasite systems are relatively stable, and there is often less variation from year to year in parasite numbers than would be expected from the variation in factors that might themselves cause variation. As a consequence, catastrophic epizootics are less common than steadier lower levels of infection and conversely effects of control measures are frequently less dramatic than anticipated.

Based on their studies on the population biology of *Camallanus oxycephalus*, Stromberg and Crites (1975) have suggested that environmental fluctuations have less effect on parasite abundance in host-parasite systems with numerous host species and several transmission pathways than on more specialized systems in which pathways are restricted. We might hypothesize, then, that as species evolve specialized mechanisms facilitating transfer from host to host, such as altered intermediate host behavior as described by Bethel and Holmes (1973); highly colored, pulsating brood-sacs of *Leucochloridium* sp. sporocysts in snail tentacles; and the *Dicrocoelium dendriticum* Hirnwürmer of Hohorst and Lämmler (1962), they restrict transmission pathways available to them, necessitating more effective regulation of numbers than provided by the environmentally labile high fecundity-high mortality system.

Through the years, knowledge of these more precise

regulatory mechanisms has resulted primarily from studies of immunology and inter- and intraspecific competition. We are blessed with a rich literature on host-parasite immunology and know much about the "crowding effect" and other aspects of competition among parasites. Recently, however, study of parasite population phenomena as regulatory mechanisms has received increasing attention. Crofton's demonstration (1971) that parasites' aggregrated frequency distributions may function as a regulatory mechanism and his subsequent definition of parasitism itself in those terms provided impetus in the immediate past.

Bradley (1974) grouped regulatory mechanisms into three general categories. His Type I, populations determined by transmission, recognizes that a degree of control is exerted by variation in transmission rates. This type of regulation, in which high fecundity compensates for transmission loss, is partially density-independent and is affected by factors extrinsic to the parasite, for example climatic conditions, but does contain density-dependent aspects.

Regulation of parasite populations at the host population level has been designated Type II regulation by Bradley (1974). Major components of Type II regulation are parasite density-dependent host mortality and some aspects of host immunity. Many parasite species demonstrate a highly aggregrated frequency distribution within their host populations (Li and Hsü, 1951; Pennycuick, 1971; and Schmid and Robinson, 1972. Crofton (1971) demonstrated how parasites with aggregrated distributions exert control over their own and host population densities through mortality in the relatively few heavily infected hosts.

Type II regulation occurs also when certain individuals of a host population mount successful immune responses resulting in destruction of their parasites. Such local exterminations result in regulation at the host population level when they occur repeatedly throughout portions of the population. Wassom, Guss, and Grundmann (1973) demonstrated this type of regulation in natural *Hymenolepis citelli-Peromyscus maniculatus* populations. Most infected deer mice displayed a protective resistance which resulted in elimination of tapeworms before proglottid maturation. Such rodents remained resistant to reinfection. Some hosts, however, were incapable of this response and retained initial and subsequent infections. Buckner and Nickol (1975) reported a similar relationship between *Moniliformis moniliformis* in *Rattus norvegicus* and between *M. clarki* in *Spermophilus tridecemlineatus*. Considerable individual variation in susceptibility of rats and ground squirrels to their respective species occurred. Under laboratory conditions some previously uninfected individuals could not be infected while 100% of the cystacanths from the same pool, but fed to other individuals of the same sex and age, were present at necropsy.

Bradley's Type III regulation designates regulation of parasite numbers by host individuals. Such regulation can occur through inhibition of parasite reproduction once a moderate density has been reached, through immunity to superinfection while the initial infection persists, and through inter- and intraspecific competition. Bradley (1974) discussed specific cases in which the first two of these Type III mechanisms may operate and parasitologists have a long history of interest in competition among parasites. Ackert, et al. (1931) demonstrated that the percentage of ascarid larvae capable of surviving in chickens decreased as the number of eggs fed increased. When varied numbers of eggs were fed in large doses, the number of worms recovered at necropsy was approximately uniform. Cross (1934) described an inverse relationship between the numbers of *Proteocephalus exiqus* and an acanthocephalan that could be found in ciscoes. Competition studies have continued to accumulate since these two early works, but mechanisms involved in the relationships are largely unknown.

Density-dependent immune sources would appear to be the most effective example of Type III regulation. Although Bradley (1974) cited an example of such regulation and Gold, Rosen, and Weller (1969) demonstrated a correlation between increasing antigenic concentrations and increased worm burdens in hamsters, quantitative data to demonstrate density-dependent immune responses are largely lacking.

Whether or not poikilothermic hosts are capable of mounting an immune response to parasites is an unresolved issue. Harris (1970; 1972) reported that chub, *Leuciscus cephalus*, produced a precipitating antibody in response to parasitism by *Pomphorhynchus laevis*, but did not demonstrate that it played a role in resistance to the parasite. In fact, the precipitin was produced only in a host species in which the acanthocephalan reached maturity and not in any of three other species in which the acanthocephalan occurred but did not mature. Later in this symposium, Kennedy will report evidence for regulation of *P. laevis* in piscine hosts, but the responsible mechanism is still unknown.

Holmes will demonstrate Bradley's Type III regulation of *Metechinorhynchus salmonis* in a piscine host. In view of Harris' (1972) finding that *P. laevis* stimulates antibody production only in a host in which it can mature, it is of particular interest that Holmes found Type III regulation of *M. salmonis* in whitefish, the primary host in his system, and not in other infected species.

In addition to the subject at hand, the five papers presented in this symposium vividly illustrate two themes in our general approach to biology. One is the complementary nature of laboratory work, field studies, and mathematical modeling. These approaches are shown to supplement one another in a manner which makes possible insights that cannot be gained by any one approach alone.

The other is the clarity with which it is demonstrated again that although there are unique aspects of parasitism, parasites are not really different from other organisms faced with making their way in nature. There is an occasional tendency to consider parasites exceptional and the host-parasite relationship different from general ecological principles. Sampling difficulties, restricted opportunities for direct observation, and complex life histories are frequently among reasons cited for vagueness in knowledge of parasite ecology. When one examines the subject matter of many recent parasitological investigations, including that of this symposium, however, it is clear that parasitologists are playing a major role in shaping concepts of population biology.

ACKNOWLEDGEMENTS

Participation assisted by funds from NIH Biomedical Sciences Support Grant RR-07055-10.

LITERATURE CITED

ACKERT, J. E., G. L. GRAHM, L. O. NOLF, and D. A. PORTER. 1931. Quantitative studies on the administration of variable numbers of nematode eggs (*Ascaridia lineata*) to chickens. *Trans. Amer. Micros. Soc. 50:* 206-214.

BETHEL, W. M., and J. C. HOLMES. 1973. Altered evasive behavior and responses to light in amphipods harboring acanthocephalan cystacanths. *J. Parasitol. 59:* 945-956.

BRADLEY, D. J. 1974. Stability in host-parasite systems. *In:* M. B. Usher and M. H. Williamson (eds.), Ecological Stability. Halsted Press, New York.

BUCKNER, S. C., and B. B. NICKOL. 1975. Host specificity and lack of hybridization of *Moniliformis clarki* (Ward 1917) Chandler 1921 and *Moniliformis moniliformis* (Bremser 1811) Travassos 1915. *J. Parasitol. 61:* 991-995.

CROFTON, H. D. 1971. A quantitative approach to parasitism. *Parasitology 62:* 179-193.

CROMPTON, D. W. T., S. ARNOLD, and D. BARNARD. 1972. The patent period and production of eggs of *Moniliformis dubius* (Acanthocephala) in the small intestine of male rats. *Internat. J. Parasitol. 2:* 319-326.

CROMPTON, D. W. T., and P. J. WHITFIELD. 1968. The course of infection and egg production of *Polymorphus minutus* (Acanthocephala) in domestic ducks. *Parasitology 58:* 231-246.

CROSS, S. X. 1934. A probable case of non-specific immunity between two parasites of ciscoes of the Trout Lake region of northern Wisconsin. *J. Parasitol. 20:* 244-245.

DEGIUSTI, D. L. 1949. The life cycle of *Leptorhynchoides thecatus* (Linton), an acantho-

cephalan of fish. *J. Parasitol. 35:* 437-460.
DEN BOER, P. J. 1968. Spreading of risk and stabilization of animal numbers. *Act. Biotheoret. 18:* 165-194.
GOLD, R., F. S. ROSEN, and T. H. WELLER. 1969. A specific circulating antigen in hamsters infected with *Schistosoma mansoni*. Detection of antigen in serum and urine and correlation between antigenic concentration and worm burden. *Amer. Trop. Med. Hyg. 18:* 545-552.
HARRIS, J. E. 1970. Precipitin production by chub (*Leuciscus cephalus*) to an intestinal helminth. *J. Parasitol. 56:* 1035.
HARRIS, J. E. 1972. The immune response of a cyprinid fish to infections of the acanthocephalan *Pomphorhynchus laevis*. *Internat. J. Parasitol. 2:* 459-469.
HOHORST, W., and G. LÄMMLER. 1962. Experimentelle dicrocoeliose-studien. *Z. Tropenmed. Parasitol. 13:* 377-397.
KATES, K. C. 1944. Some observations on some experimental infections of pigs with the thorny-headed worm, *Macracanthorhynchus hirudinaceus*. *Amer. J. Vet. Res. 5:* 166-172.
LI, S. Y., and H. F. HSÜ. 1951. On the frequency distribution of parasitic helminths in their naturally infected hosts. *J. Parasitol. 37:* 32-41.
PENNYCUICK, L. 1971. Frequency distributions of parasites in a population of three-spined sticklebacks, *Gasterosteus aculeatus* L., with particular reference to the negative binomial distribution. *Parasitology 63:* 389-406.
SCHMID, W. D., and E.J. ROBINSON, JR. 1972. The pattern of a host-parasite distribution. *J. Parasitol. 57:* 907-910.
STROMBERG, P. C., and J. L. CRITES. 1975. Population biology of *Camallanus oxycephalus* Ward and Magath, 1916 (Nematoda: Camallanidae) in white bass in western Lake Erie. *J. Parasitol. 61:* 123-132.
WASSOM, D. L., V. M. GUSS, and A. W. GRUNDMANN. 1973. Host resistance in a natural host-parasite system. Resistance to *Hymenolepis citelli* by *Peromyscus maniculatus*. *J. Parasitol. 59:* 117-121.

Parasitism and r- and K-selection

GERALD W. ESCH, TERRY C. HAZEN AND JOHN M. AHO

Department of Biology
Wake Forest University
Winston-Salem, North Carolina
U.S.A.

and

Savannah River Ecology Laboratory
Aiken, South Carolina
U.S.A.

INTRODUCTION

   The relationship between an organism and its environment can be studied in a number of ways. But for the most part, since suggested by Elton in 1927, the essence of ecology has been the examination of the distribution and abundance of plants and animals in nature. Such an approach can be made at the individual, population and ecosystem levels of biological organization.
   While the accent in parasitology has largely been at the individual level and, hence, an emphasis on physiology and immunology, in recent times there has been an increasing interest in studying parasitism at the population and ecosystem levels. Andrewartha (1970) has said that the aim of population ecology is to explain the numbers of plants or animals which can be counted or estimated in natural populations. In seeking an explanation for the numbers of individual organisms, para-

sitic or otherwise, there are at least two problems
which must be resolved. First, the numbers of organisms which may be present at a given point in time must
be counted, or estimated, accurately. And second, the
fundamental nature of change in numbers should be perceived, analyzed and related, if possible, to biotic
and abiotic variables. In other words, changes in density in a temporal sense must be ascertained and then,
an effort must be made to identify the factors which
regulate or control the density changes. Obviously,
there are a number of highly complex, contributing
variables, but in many instances, these have been more
than adequately defined and the result has been the
formulation of meaningful interpretations, applicable
to the regulation of population dynamics.

Many of the difficulties inherent in understanding
the regulation of free-living populations are also
critical in the consideration of parasite populations.
However, the very nature of parasitism inexorably adds
a diversity of problems unique to parasitic organisms
and thus makes an understanding of regulatory phenomena
even more difficult. Most significant among these
problems is that the immediate environment of the parasite is alive and thus, potentially, is capable of responding in a negative manner such that further recruitment and establishment may be restricted. Conversely, other parasites are able, under certain circumstances, to induce host mortality, thereby elimi-
nating their environment and preventing further recruitment. It should be clear, therefore, that the
complexity of host-parasite interactions may pose very
serious obstacles for understanding the nature of regulatory phenomena associated with parasite populations.

Prior to considering the host-parasite relationship and the regulation of parasite populations, a
problem of immediate concern is to precisely define
what it is that constitutes a parasite population.
The definition is of importance since, if regulation
is to be understood, suitable operational terminology
must be employed. For example, it would generally be
acceptable to define a population of free-living
plants or animals as a group of organisms of the same
species occupying a given space. As pointed out by
Esch, Gibbons, and Bourque (1975), however, this defi-

nition is unsatisfactory for parasite populations. The problem, as represented by these authors, is whether "all members of a given parasite species within a single host constitute a population, or should all members of a species in all hosts within a given ecosystem be considered a population?" It also was noted that populations of free-living organisms increase in density through birth and/or immigration processes while, within a single host, the density of most species of helminth parasites is able to increase only through immigration (= recruitment).

In an effort to circumvent the difficulties attendant in defining parasite populations in the same terms as free-living populations, a new approach was devised by Esch, Gibbons, and Bourque (1975). They proposed that all individuals of a single parasite species within an individual host be regarded as an infrapopulation. In broader terms, they referred to all individuals of a given parasite species, in all stages of development, within all hosts of an ecosystem as a suprapopulation.

The justification for a new approach becomes more apparent when a search is made of the current literature dealing with the ecology of animal parasites. There has been a strong tendency, for example, to treat parasite population biology in terms of specific, isolated life cycle stages, without regard for other life cycle stages of the same species which may also be present within the same ecosystem. The result has been the generation of considerable, and generally useful, data which are essential for analysis of infrapopulation dynamics. It does not, however, provide information necessary for evaluating the dynamics of the suprapopulation. Exceptions to the more widely used approach are the elegant studies by Hairston (1965), who examined the population biology of the trematode, *Schistosoma japonicum*, at the suprapopulation level and Anderson (1974), who also used a more holistic approach in studying the population biology of the tapeworm *Caryophyllaeus laticeps*. The utility of these approaches and that of Esch, Gibbons, and Bourque (1975) should be obvious. It would not serve any useful purpose to study the population biology of a free-living species by examination of only the adults or only the juveniles. It is important that knowledge of all life cycle stages, free-living or otherwise, be obtained and assessed if an understanding of regulatory phenomena is to be gained. When individual life cycle stages, and therefore infrapopulations, are studied separately and information concerning each stage is then summarized and integrated, only then will it be possible to understand regulation at both the infra- and suprapopulation levels.

A useful method for considering the regulation of parasite infra- and suprapopulations was devised by Kennedy (1970). He formulated a "general systems theory" which he believed would assist in the analysis and understanding of those factors which are important in regulating parasite systems. While his efforts were more specifically directed at identifying the biotic and abiotic forces which operate in fish host-parasite systems, they are nonetheless useful in examining regulation in broader terms. Borrowing from Kennedy (1970) and adding the notions presented by Esch, Gibbons, and Bourque (1975), a simplified, yet general, description of regulation as applied to the host-parasite system, has been developed (see Figure 1).

INTRINSIC (TO THE HOST) VARIABLES AFFECTING REGULATION OF PARASITE INFRAPOPULATIONS

In order to perceive the processes involved in regulation of parasite infrapopulations, it is necessary to focus on the flow of parasites (Figure 1) into (recruitment) and out of (turnover) the host. When this is done, then it can be seen that the density of an infrapopulation is a function of several intrinsic (to the host) variables.

INHERENT HOST ACCEPTABILITY

The emphasis is on inherent, for it infers a genetically-based characteristic(s) which ultimately determines whether a parasite will become established once it has been recruited. Such factors are of obvious significance since they are directly related to the

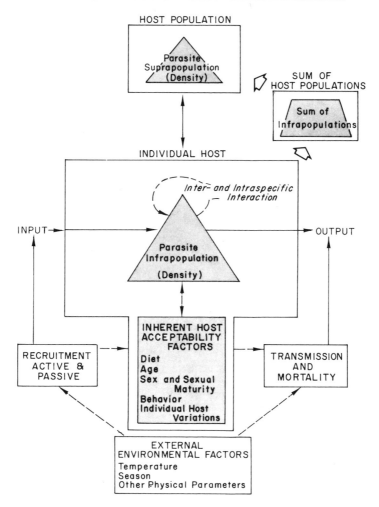

Figure 1. *A schematic model showing the relationship between inherent host acceptability factors and external environmental factors in regulating parasite input and output at the infrapopulation and suprapopulation levels.*

phenomenon of host specificity. Odening (1976) lists nine different criteria which may be useful in measuring the suitability of a given host for a given parasite and several of these are of significance in determing inherent host acceptability. Included among these are the rate of parasite development, life expectancy of the parasite (and the host), reproductive potential of the parasite, intensity of the host's defensive response, ability of the host to withstand pathogenic insult, and the viability and transmissability of the reproductive forms of the parasite. Collectively, these criteria will be reflected in host and parasite compatability (or incompatability) and ultimately are the manifestations of a variety of morphological and physiological characteristics.

For each species of organism, parasitic or otherwise, there is an array of requirements which, in combination, represent a multidimensional space, or hypervolume. The physiological and morphological requirements of a parasitic organism, relative to all the biotic and abiotic factors which influence the host and parasite, represent the niche. For parasitic organisms occupying the intestine of a host, the requirements include a range of morphological, physical and biochemical factors. Smyth and Smyth (1968) reported, for example, that the variability in sizes of the crypts of Lieberkuhn could be of importance in the establishment of *Echinococcus granulosus* and *E. multilocularis* in various species of carnivores. Should the crypts be too narrow or shallow, then one of the species could not become established (even after recruitment), while the other exhibited a wider range of potential hosts since its establishment was not limited by the size of the crypts. An even better example was reported by Williams (1960). Thus, the fish, *Raja montagui*, *R. clavata* and *R. naevus*, have intestines with substantial morphological variability. Each fish species is parasitized by species of the tetraphyllidean tapeworm, *Echeneibothrium*, and, in each case, the cestode scolex is appropriately modified to conform with variations in villus depth as well as morphology of the crypts.

Bile salts within the intestines of vertebrate species are known to vary both in composition and concentration. "Since a wide range of bile salts occurs in vertebrates, it may be expected that the composition of bile - along with other factors - may determine whether or not a particular species can develop in a particular host" (Smyth, 1969). For example, it was shown by Smyth (1962) that the tegument of *Echinococcus granulosus* is lysed by deoxycholic acid as well as by bile from several herbivorous, but 'unsuitable hosts', such as rabbit, sheep or hare, which have high concentrations of deoxycholic acid. On the other hand, bile from carnivorous animals such as fox, dog and cat, have relatively little effect since deoxycholic acid is virtually non-existent.

There are numerous other cases which could be used as examples to illustrate the notion of 'inherent' host acceptability following recruitment of a parasite. It is clear enough from these, however, that host and parasite morphology and physiology have evolved in such a way that compatability (or incompatability) with a specific host, or group of hosts, will be the outcome of recruitment. While the parasite has evolved morphologically and physiologically to conform to a prescribed set of host-provided environmental conditions, the host also has evolved in such a manner that it is capable of responding to recruitment of parasites. In general, this adaptation by a host to a parasite produces incompatability, or at least partially so. The thrust of this evolutionary process is defensive in nature and has resulted in the development of an immune capability which, unless reversed by the parasite evolving a counter strategy, e.g. molecular minicry (Damian, 1964), non-reciprocal cross-immunity (Schad, 1966), should produce mortality of the parasite or, at the least, make it more difficult for establishment once recruitment has occurred. The phylogenetic and ontogenetic consequences of immunity thus may be reflected either in compatability or incompatability of host-parasite systems.

HOST FACTORS

Inherent host acceptability could be considered as the single, basic, controlling element (excluding, of course, external environmental forces) in influencing recruitment and establishment of parasite infrapopulations. There are, however, several additional factors which can be considered independently even though they are actually modifiers of genetically based strategies. These factors include dietary considerations, host age, sexual maturation and behavior.

Diet

It can be said that, in general, the enteric parasite fauna of a host is a reflection of what is eaten. Moreover, as pointed out by Dogiel (1964), "the parasitological indicators of diet are among those clues which allow us to make deductions from the type of parasite fauna about various aspects of the ecology of the host."

There are numerous studies which have adequately shown the relationship between diet and the parasite fauna. Dogiel (1964), for example, noted that the parasite fauna of the herbivorous cyprinid, *Chondrostomus nasus*, was virtually devoid of enteric helminths while carnivorous cyprinids, within the same ecosystem had relatively 'rich' enteric faunas. Esch and Gibbons (1967) reported that sexually immature (carnivorous), painted turtles, *Chrysemys picta*, were much more heavily parasitized than herbivorous adults. In the case of both cyprinid fish and painted turtles, the obvious variations in parasite fauna were the result of differences in probability of ingestion of infected intermediate hosts.

In lacustrine ecosystems, ecological succession may proceed from oligotrophy through mesotrophy to a eutrophic condition. As might be expected, the successional changes are reflected in complex alterations in the biotic community. Wisniewski (1958), Chubb (1970) and Esch (1971) have all reported on the differences between oligotrophic and eutrophic ecosystems with respect to the diversity and density of the parasite fauna within the piscine community. Esch (1971) proposed that the nature of the changes during succession could be explained by examining the changes in predator-prey interaction. Thus, in an oligotrophic lake, the majority of parasites complete their life cycles within interacting species of the aquatic fauna; there is little aquatic-terrestrial interaction and, hence, the system is relatively closed in terms of predator-prey relationships. On the other hand, the parasite fauna of the eutrophic ecosystem reflects more extensive aquatic-terrestrial interaction with more parasites of fish completing their life cycles in terrestrial predators; the eutrophic system is thus more open. As the biotic community changes during succession, the predator-prey interactions also shift, resulting in both qualitative and quantitative changes in the parasite fauna. Support for this hypothesis was recently provided by Kennedy (1975a).

As proposed by Esch (1971), therefore, "the nature of predator-prey relationships should serve...as a potential biological index for predicting the structure of a parasite fauna in any given aquatic ecosystem." While this approach was directed mostly at long-term successional changes, the same principles should also apply to seasonal alterations in the biotic community within aquatic ecosystems. There are a number of typical transformations which occur on a seasonal basis in the biotic and physicochemical characteristics of (especially) temperate zone aquatic ecosystems. Many of these changes ultimately can be attributed to increase in temperature and light intensity; there is likewise a shift in quantity and quality of both producer and consumer species. It naturally follows that the dietary components of many consumer organisms will also shift. Concomitant seasonal changes in the density and diversity of the parasite fauna within various hosts should also occur and, in many cases where studied, such has happened. Conner (1953) reported a significant seasonal change in density of the tapeworm *Proteocephalus stizostethi* in pikefish, *Stizostedion v. vitreum*. He stated that tapeworms were variable in size in October and November and that recruitment did not occur later in winter. He indicated that the variability in size during fall was due to intermittent feeding on the intermediate host (presumably a cyclopoid copepod) and that the lack of recruitment in the winter was occasioned by the unavailability of the intermediate host in the diet. Lees (1962) regarded the seasonal changes in frog trematode density and diversity as being due to the seasonal changes in abundance of various insect intermediate hosts. Gerking (1962) reported that the diet of bluegill sunfish, *Lepomis macrochirus*, in an Indiana (U.S.A.) eutrophic lake, turned from a high proportion of *Daphnia* in July to a high proportion of midge larvae in August. Esch, et al. (1976) reported a significant shift in the recruitment of the acanthocephalans *Leptorhynchoides thecatus* and *Pomporhynchus bulbocolli* during an eight week period in the summer of 1973 and attributed the variation in recruitment to changes in diet.

Rysavy (1966) examined the cestode fauna of birds representing several different orders, as well as different food habits and habitat preferences. The study indicated that similarities in food and habitat preferences may result in similar or even identical cestode faunas even if the birds are phylogenetically widely separated. Conversely, species of birds which are closely related phylogenetically may have very diverse faunas if their diet and habitat preferences are widely divergent.

Age

Host age is an important factor in considering regulation of parasite infrapopulations. The patterns of change in parasite densities and diversity, as the host ages, are quite variable. In some cases, a given species of parasite will be intensively recruited while the host is young and then decline in numbers as the host ages. In others, the timing of recruitment will be reversed, with the parasite being absent in young individuals and appearing with greater frequency as the host ages. In yet another pattern, recruitment will begin while the host is a juvenile and persist throughout its lifetime. Changes in quality and quantity of the parasite fauna may be a consequence of several interacting variables, including diet, immunity, interspecific interaction among sympatric species and changes in foraging patterns as the host ages.

Lewis (1968) studied the changes in the parasite fauna of the long-tailed field mouse, *Apodemus s. sylvaticus*, in two contrasting habitats from Skomer Island, off the coast of England. Results indicated that *Nematospiroides dubius*, a nematode parasite not requiring an intermediate host, was maintained in adult hosts throughout the 15 month study period. Juvenile *A. sylvaticus* became infected during and after the breeding season in the summer and were responsible for maintaining the infection cycle into the following year when they themselves became adults. According to Lewis, "a striking feature here, however, is that the juveniles are less likely to be infected than the adults, a fact which is correlated with the life history of *N. dubius* and the limited foraging activities of the juveniles during and immediately after the breeding season." Since eggs of *N. dubius* are shed with feces, the density of the infective larvae "will obviously depend upon the size of the area over which the host moves, whether the host feces are distributed at random or are aggregated, a larger area of host movement will give rise to a low density of infective stages, resulting in a low level of infestation in the host. Even if the host's movements range over a large or small area, the adults due to their increased foraging activity, will probably harbor heavier worm burdens of *N. dubius* than the juveniles."

Hine and Kennedy (1974) reported a situation in which the acanthocephalan *Pomphorhynchus laevis* was more or less constantly recruited throughout the life of the definitive host, *Leuciscus leuciscus*. There was also an increase in the infrapopulation density of *P. laevis* as the host aged. Since, however, the parasite infrapopulation was found to be in a state of virtual dynamic equilibrium, i.e. constant recruitment and turnover, the increasing densities with age could not be attributed to the addition of newly recruited parasites to a previously established infrapopulation. Instead, the change in density with age was attributed simply to an increase in the level of feeding with increased age.

The allocreadid trematodes *Bunodera sacculata* and *B. luciopercae* in yellow perch, *Perca flavescens*, in Lake Opeongo, Ontario, Canada, were both reported to increase in infection percentage as the host aged (Cannon, 1972). In contrast to *P. laevis* which were in dynamic equilibrium (Hine and Kennedy, 1974), both allocreadid species were recruited and turned over on an annual basis. Presumably, greater recruitment in older fish was the result of higher probability of exposure through wider foraging or heavier feeding, or both.

Nelson (1959) reported that *Schistosoma haematobium* was primarily an infection of children and that evidence of infection in adults is difficult to demonstrate. Clarke (1965), as reviewed by Kagan (1966), indicated that, with increasing age, there was a decrease in the level of infection accompanied by a decline in egg production. The pattern was attributed to an increasing level of acquired immunity in the population under consideration. Age related immunity to several other species of parasitic helminth has been reported (Horak, 1971; Soulsby, 1963; Kassai and Aitken, 1967; Michel, 1969). In the majority of cases, immunity is acquired through exposure to infective larval stages. In this way, as the potential host ages, the probability of exposure increases but immunity also increases. The outcome is heavier infections in young-age groups and a decline in older ones. As pointed out by Michel (1969), "population increase of a parasite which completes its life cycle within the same host tends to follow a constant pattern. Characteristically, the increase is exponential during early stages of infection while the host offers an ideal environment. Subsequently, when the host becomes resistant and represents a less suitable environment, the rate of increase declines to zero and the population then rapidly decreases."

Sex and Sexual Maturity

In some host-parasite systems there may be a strong correlation between the onset of sexual maturity and the density and diversity of the parasite faunas; with other systems, there appears to be a relationship between host sex and the intensity of infection. In either situation, the relationship between sex and parasitism could be due to variation in recruitment by males and females, or it may be due to the quantity or quality of steroid hormones (presumably, mainly androgens and estrogens) which differentially affect establishment of parasites after recruitment. The relationship between host sex and the establishment and maintenance of parasites has a long history of documentation although, in virtually all cases, there is a complete lack of evidence to indicate that such a correlation is *directly* causal.

Studies by Addis (1946), Beck (1952), Esch, (1967), Culbreth, Esch, and Kuhn (1972) and Novak (1975) have shown that the sex of the host can variously affect establishment, growth, egg production or asexual reproduction in an array of helminth parasites. Fischer and Freeman (1969) reported that the plerocercoids of the tapeworm *Proteocephalus ambloplitis* are stimulated to migrate from parenteric sites into the intestine of the smallmouth bass, *Micropterus dolomieui*. The cue for migration was suggested to be rising water temperature. Interestingly, however, they also reported that migration, and subsequent maturation to the adult stage, did not occur unless bass were sexually mature. Their findings were later confirmed by Esch, Johnson, and Coggins (1975). The implication of these studies is clear, i.e. sexual maturity coincides with increased production of sex hormones which may well be involved in stimulating migration, or subsequent growth and development of the tapeworm. It should again be emphasized, however, that evidence to support a direct causal relationship between steroids and recruitment and establishment of parasites in any vertebrate host is lacking.

Behavior

Host behavior may play a very significant role in the regulation of parasite population dynamics. The role is manifested to the greatest extent by influence on recruitment, rather than establishment, once exposure has occurred. Any behavioral characteristic which ensures, or prevents, contact between a host and a parasite would thus be of importance. It should also be emphasized that the behavioral attributes of either the host or the parasite may be of significance. The literature in this area is replete with examples and thus only a few will be mentioned here.

One of the best studies on the behavior of a freeliving parasite larval stage was conducted by Donges (1963). By varying light intensity and water turbulence it was shown that the swimming behavior of the cercaria of *Posthodiplostomum cuticola* could be significantly altered. When cercaria were maintained under relatively stable light intensity and then, suddenly, light intensity is reduced to zero, followed by restoration to the original level, there was a sig-

nificant increase in swimming activity. On the other-
hand, a linear reduction in light intensity over a two
second interval produced complete inhibition of swimming activity. It was suggested that the sudden drop
in light intensity simulated a shadow produced by a
swimming fish and that the increased swimming activity
of the cercaria would increase the probability of contact. Similarly, water turbulence was said to simulate
currents created by a swimming fish and also thus increase the chance of contact between the fish and parasite. A slow decrease in light intensity produced no
response. It was suggested that the slow decrease in
light intensity would be similar to that which might
occur with changing cloud cover. Biologically, the
significance of the cercarial swimming behavior would
rest in the preservation of limited glycogen reserves
in the non-feeding larval stage.

In a recent review, Holmes and Bethel (1972) described a number of host-parasite systems in which the
host behavior was modified to such an extent that the
probability of parasite transmission to a subsequent
host in the life cycle was significantly increased.
They detailed a series of studies in which the behavior
of the amphipod, *Gammarus lacustris*, was altered when
infected with the cystacanth larval stage of the acanthocephalan, *Polymorphus paradoxus*. Thus, when infected, *G. lacustris* exhibited a striking, positive photo-
taxis in conjunction with a strong clinging behavior
such that they tightly attached to floating vegetation,
remaining even when vigorously agitated. The consequence of the abnormal response was that infected *G.
lacustris* became vulnerable to predation by mallard
ducks, the definitive host for the parasite. According to Holmes and Bethel (1972), "the combination of
the different behavior patterns of infected and uninfected gammarids, and the feeding behavior of mallards,
resulted in a disproportionately large number of infected gammarids being eaten" during the course of a
series of prey selection experiments.

A number of other cases of behavioral modification
as a result of parasitism were also reviewed by Holmes
and Bethel (1972). Thus, van Dobben (1952) reported
that approximately 6.5% of a population of roach,
Rutilis rutilis, were infected with plerocercoids of

the tapeworm, *Ligula intestinalis*, while 30% of the roach in the diet of the cormorant definitive host were infected; this would imply that infected roach were more vulnerable to predation. Lester (1971), suggested that the behavior of sticklebacks infected by plerocercoids of the tapeworm *Schistocephalus solidus* was altered in such a way that predation by piscivorous birds was increased. Previous studies by Arme and Owen (1967) had shown that the gonadal-hypophyseal hormonal axis was altered by a secretion from the plerocercoids. In effect, the fish became stunted and reproductively sterile while the larval tapeworm continued growing, ultimately occupying substantial space in the body cavity thereby affecting the ability of the fish to swim in an upright manner; the outcome was increased predation by the fish-eating definitive host.

A recent review by Croll (1975) and earlier ones by Rogers and Summerville (1963) and Michel (1969) have most adequately described various features of the behavior of larval nematodes, particularly in terms which would apply to the regulation of parasite populations. In general, it can be stated that nematode larvae exhibit responsive capabilities to a large number of chemical, mechanical, thermal and light stimuli. In being able to respond to this array of stimuli, nematode larvae are able to seek, penetrate and establish within appropriate intermediate and definitive hosts, thereby contributing to the regulation of infrapopulations.

INTER- AND INTRASPECIFIC INTERACTIONS

The density (and diversity) of parasite infrapopulations may be affected by inter- and intraspecific interactions. Thus, several investigators (Holmes, 1961; Schad, 1963; Chappel, 1969), studying quite different host-parasite systems under both laboratory and field conditions, have shown that enteric parasite populations can be affected by competition for space or nutrients (additional comments regarding these studies and other related ones will be made subsequently).

EXTRINSIC (TO THE HOST) FACTORS AFFECTING THE REGULATION OF PARASITE INFRAPOPULATIONS

Input and output from the infrapopulation are the manifestations of recruitment, transmission and mortality (refer to Figure 1). Recruitment may be characterized as being active or passive. Active recruitment would require an expenditure of energy on the part of the host, i.e. energy expended by a predator in seeking prey which may harbor an infective larval stage for the definitive (predator) host. In other cases, active recruitment would occur when a potential host accidentally ingests an egg or larva along with other food material (there would still be an expenditure of energy by the host in recruitment of the parasite). Passive recruitment would occur when a parasite expends energy in order to seek and penetrate a host; in this case, the host is passively involved in recruitment.

Output from an infrapopulation may result from parasite mortality, or it may occur when an infrapopulation is transmitted to a new host either by direct predator-prey interaction or via direct inoculation of a parasite by an insect vector. Mortality among parasite populations may be the result of normal senescence which can occur seasonally, or at irregular intervals. Mortality may also be a reflection of host and parasite incompatability such as might be induced through immunologic means. A change in diet by a host can have profound affects on enteric physiology and may, as a result, create local environmental conditions which are incompatible with continued existence of an established parasite infrapopulation - the result would be parasite mortality and output. Transmission may also affect the density of an infrapopulation. The density may decline partially, or completely, depending on the specific nature of the parasite's life cycle. For example, the host might be serving as an intermediate for the parasite which must be consumed by the appropriate definitive host in order to complete its life cycle. Obviously, the outcome of this kind of transmission is a decrease in the infrapopulation density. These kinds of changes would be of significance when the sum of similar infrapopulations is under consideration, or when the density of subsequent life history stages and infrapopulations is considered, or when the dynamics of the suprapopulation is the focus of attention.

Recruitment (whether active or passive), as well as mortality and transmission, are influenced by an array of abiotic and biotic factors; any variable impacting on the physiology and behavior of either the host or the parasite, or both, may be important in regulation. As stated by Esch, Johnson, and Coggins (1975), "the result of the collage of interacting host and environmental variables is the establishment of isolated parasite infrapopulations which collectively represent an important segment of the suprapopulation." The number of these factors is large and would be cumbersome either to list or thoroughly comment upon, so the following are noted as examples only.

TEMPERATURE

In general, the effect of temperature is measured in terms of its influence on the rate of enzymatically catalyzed metabolic activities, but it also affects such non-catalytic activities as diffusion and osmosis. Operating within this framework, temperature affects a wide variety of growth and development processes, any one of which may effectively influence recruitment, transmission and mortality, and thus be involved with regulation of parasite populations. In some cases, addition of thermal effluent may exacerbate what may be considered as normal temperature effects.

Not surprisingly, temperature and its role in controlling egg hatching, development of intramolluscan trematode stages, and the subsequent release of cercaria are all known to affect the regulation of trematode populations. In an excellent review, Ollerenshaw and Smith (1969) described the use of meteorological factors in the forecasting of helminthic disease in a number of animals, for example, fascioliasis in sheep. Employing observations by Ollerenshaw (1959), they described "a time-table showing the seasonal distribution of eggs on pasture, infected snails, infected herbage and disease in sheep...(which) was based solely on the response of the parasite to seasonal variations in temperature." Spall and Summerfelt (1970) reported that higher water temperature increases both the maturation rates of the intramolluscan larval stages of *Posthodiplostomum minimum* and the release of cercaria. Both of these developmental processes play important roles in determining rates of passive recruitment and densities of metacercarial infrapopulations within piscine intermediate hosts.

Kennedy (1975b) has correctly pointed out that there are a number of fish parasites in which egg production by the adult and the maturation cycle as a whole can be closely correlated to annual changes in the water temperature cycle. And, as indicated by Chubb (1967), the correlation suggests a causal relationship. Since, however, there is no supportive experimental evidence, Kennedy (1975b) urges caution in drawing conclusions with respect to the influence of temperature on these growth processes.

Bauer (1959) reported that the development time of the protozoan parasite of fish, *Ichthyopthirius mutifiliis*, was inversely correlated with rising temperature; and, indeed, this may be true for many protozoan parasites. Recently, Esch, et al. (1976) reported a striking seasonal periodicity in infection of largemouth bass, *Micropterus salmoides*, by the peritrich ciliate *Epistylis* sp. This periodicity of infection was very closely tied with seasonal changes in mean temperature. Seasonal changes in body condition of the host fish were, however, also reported; it was suggested that the body condition of the host may be of overriding significance in affecting the epizootiology of *Epistylis*. This observation supports the generalizations made by Kennedy (1975b) that correlations between temperature and growth and development of parasites do not necessarily indicate a direct causal relationship between temperature and changes in parasite population dynamics.

In a series of studies by Fischer (1967) and Fischer and Freeman (1969) on the biology of the tapeworms *Proteocephalus fluviatilis* and *Proteocephalus ambloplitis* in smallmouth bass, a strong case was made for a correlation of several growth factors with temperature. In the first instance, the eggs of *P. fluviatilis* were reportedly killed by low winter temperatures, and since infection is then precluded in either copepods or bass, infrapopulation densities in both hosts decline (Fischer, 1967). Fischer and Freeman (1969) implicated temperature as a force in inducing the parenteric plerocercoids of *P. ambloplitis* to migrate into the intestine, thus resulting in enteric recruitment of adult tapeworms.

In a reservoir (Par Pond, Aiken, S. C., U.S.A.) receiving thermal effluent from a nuclear production reactor, Eure and Esch (1974) reported that while infection percentages did not vary between locations, the densities of *Neoechinorhynchus cylindratus* (Acanthocephala) infrapopulations in largemouth bass, were significantly higher in bass taken from thermally altered areas as compared to those taken in ambient areas. They attributed the difference to heavier feeding activities of bass subjected to elevated temperature and not to any direct effects of temperature on the parasite. In the same reservoir system, Bourque and Esch (1974) reported a similar thermal effect with respect to *Neoechinohynchus* spp. in the yellow-bellied turtle, *Psuedemys s. scripta*. Turtles in areas receiving thermal loading had significantly higher infrapopulation densities than those taken in ambient areas. They also attributed the differences to increased feeding by the turtles. However, the diversity of the parasite fauna was less among turtles from the thermally altered areas even though densities of the remaining parasite species were high. The reduced diversity was probably due to a decline in species diversity of the ostracod intermediate host, since similar changes in density and diversity had previously been noted in other members of the biota of that reservoir (Parker, Hirshfield, and Gibbons, 1973). Aho, Gibbons, and Esch (1976), who also studied parasitism in the Par Pond system, reported a significant difference in metacercarial infrapopulations in the mosquitofish, *Gambusia affinis*, in relation to thermal effluent. Their results showed that while infection percentages were similar in *G. affinis* throughout the reservoir, fish taken in heated areas had higher worm burdens of the metacercaria of *Ornithodiplostomum ptychocheilus* than in those from ambient areas. Conversely, *G. affinis* from ambient areas of Par Pond had larger infrapopulation densities of the metacercaria of *Diplostomulum scheuringi* than did fish from thermally altered areas. Interestingly, recruitment of these parasites by *G. affinis* is passive, while recruitment in the hosts studied by Eure and Esch (1974) and Bourque and Esch (1974) was active. Clearly, further efforts are needed to evaluate the impact of thermal effluent on recruitment and turnover of parasites by poikilothermic vertebrates as well as invertebrates.

DISPERSION

The phenomenon of dispersion is exceedingly important in considering those factors involved in the regulation of parasite infrapopulations. In terms of parasitic organisms, Crofton (1971) was the first to call attention to the relationship between regulation and dispersion. Indeed, he attempted to incorporate the concept of dispersion into a comprehensive definition of parasitism. The essential features of his definition were: (1) that parasites are physiologically dependent on their hosts, (2) the parasite kills heavily infected hosts, (3) the parasite species has a higher reproductive potential than its host, and (4) the infection process produces or tends to produce an overdispersed distribution of parasites within the host population. Whether all four tenets are accepted or not is not essential to this discussion. The important feature is the last, i.e. overdispersion. Since Crofton's 1971 paper, there has been a series of reports (Pennycuick, 1971; Schmid and Robinson, 1972; Bourque and Esch, 1974; Boxshall, 1974; Anderson, 1974) which have directed attention to overdispersion as a characteristic of various parasite populations.

Indeed, as was shown by Anderson (1974), the overdispersion of infected intermediate hosts, in conjunction with feeding behavior of the definitive host, may directly influence recruitment of parasite infrapopulations and thus be of significance in regulation. He reported that the overdispersion of adults of the tapeworm *Caryophyllaeus laticeps* in the European bream could be described by a negative binomial model. He suggested that the overdispersed infrapopulation densities of the adult tapeworms were not due to the compounding of random successive waves of invasion by infective stages acquired by predation of randomly scattered infected intermediate hosts (benthic tubificid worms). Instead, he attributed the overdispersed infrapopulations to the spatial heterogeneity of infected intermediate hosts within the benthic community. The spatial distribution, when coupled with non-random feeding behavior of the bream, was sufficient to produce overdispersion of adult worms in the definitive host.

REGULATION AT THE SUPRAPOPULATION LEVEL

If successful recruitment by an appropriate host occurs for a particular life cycle stage in any given host-parasite system, then an infrapopulation may become established. It will then be maintained for an appropriate period of time after which individuals within the infrapopulation may senesce and die or the entire infrapopulation may be transmitted to the next host in the life cycle, becoming a new infrapopulation or part of an existing one. The relationships among the factors which regulate at the infrapopulation and suprapopulation levels are exceedingly difficult to analyze, in part because the complexity of the regulatory phenomena acting at these levels is frequently compounded by the existence of multiple life cycle stages involving a variety of intermediate and definitive hosts.

One of the few analyses of suprapopulation dynamics yet published examined was based on the digenetic trematode *Schistosoma japonicum* (Hairston, 1965; Pesigan, et al., 1958). The outcome of these studies showed that field rats maintained the schistosome suprapopulation at such a level that if all humans were to be removed from a given area and replaced by the same number of uninfected individuals, then only a few years would be required for infrapopulations within humans to reach previous levels. For an excellent consideration of regulation at the suprapopulation level, see Holmes, Hobbs, and Leong (this volume).

r- AND K-SELECTION

Among host-parasite systems there is a series of selection forces (biotic and abiotic) which, acting in concert with genetic variability of both the hosts and the parasites, ultimately determines the success of individuals, populations and species. Thus, selection pressure(s) and genetic variability within the host and parasite are reflected in regulation. Some of the ideas inherent in the notion of natural selection and regulation of parasite populations can be suitably joined in the concept of r- and K-selection which has produced considerable discussion, both pro and con, among ecologists in recent years. It is not the intention here to enter the controversy centering on r- and K-selection. Instead, use will be made of the concept only for the purpose of generating new, even perhaps incautious, ideas concerning the factors involved in the regulation of parasite populations. Liberal use will also be made of the notions presented in the previous section dealing with the schematic model shown in Figure 1.

The concept of r- and K-selection was initially proposed by Dobzhansky in 1950. Writing in American Scientist, Dobzhansky suggested that natural selection in the tropics favored low fecundity and slow development while, in temperate zones, competitive advantage was bestowed on those species with higher fecundity and more rapid development. Pianka (1970) pointed out that "Dobzhansky's ideas were framed in terms too specific to reach the general ecological audience and have gone more or less unnoticed until fairly recently."

The terms r-selection and K-selection were actually coined by MacArthur and Wilson (1967). Their intent was to characterize temperate and tropic zone selective mechanisms but, as Pianka (1970) indicated, these two types of selection are not at all restricted to the tropics or temperate climates. In 1970, Pianka presented a summary of the concepts of r- and K-selection. His view was that there is an r-K continuum, with the r-end point representing a situation in which the optimum strategy is to place 100% of the matter and energy of an organism into reproduction, with a minimum into each individual offspring so as to produce the maximum number of progeny (strategy in this paper is used to indicate a genetically fixed adaptation or characteristic, as in the sense of Harper, 1967). The outcome of r-selection would thus be high productivity. Continuing, Pianka (1970) indicated that at the other end of the continuum, K-selection would lead to highly efficient utilization of environmental resources. Density effects would be maximized and the environment would be saturated with organisms; in other words, the population would be at capacity for a given amount of space or nutrient level. Under such conditions, competition would be keen, with the best survival strategy being to shunt all available resources into self-maintenance and to produce highly fit progeny. As McNaughton (1975) stated, "in the ecological void, the optimal adaptive strategy channels all possible resources into progeny, thereby maximizing the rate at which resources are colonized. At ecological saturation, on the other hand, the optimal strategy channels all possible resources into survival and production of a few offspring with highly competitive ability."

The population growth curve for a K-strategist can be written in the form shown in Figure 2 (top). The

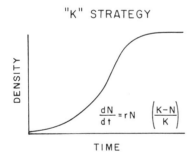
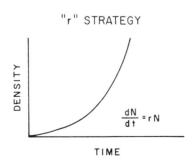

Figure 2. Population growth curve for a K-strategist (at the top) and an r-strategist (at the bottom).

curve can be described by the familiar Verhust-Pearl logistic equation. The equation, in essence, says the following: the rate of population increase ($\frac{dN}{dt}$) is equal to the maximum growth rate times the number in the population (rN), times the degree of realization of maximum rate ($\frac{K-N}{K}$) (Odum, 1971). In this equation, K represents the upper asymptote, or the carrying capacity; K-selection relates to competitive ability or, in other words, the manner in which the instantaneous increase of a population will change in relation to changing levels of competition.

The population growth curve for an r-strategist will be recognized in Figure 2 (bottom) as the expotential phase of the previous logistic equation where a population is increasing (or decreasing) expotentially. The r-term in the equation represents what is commonly called the intrinsic rate of natural increase; it is a constant, but it may take any value. According to Hariston, Tinkle and Wilbur (1970), "it thus represents in a single number all of the physiological response of all members of the population to a given set of environmental conditions since, ultimately, all physiological responses must be relatable to the ability to reproduce, or the ability to survive."

As previously indicated, the concept of r- and K-selection has not been universally accepted (for a discussion of some objections, see Hariston, Tinkle, and Wilbur, 1970). In spite of their criticisms (which may or may not be well taken) there is obvious significance in its acceptance in general terms and, especially, in the sense presented by Pianka (1970; 1972). It has gained wide acceptance by a variety of investigators because it does focus on an ecologically important problem, i.e. "allocation of resources between competitive and reproductive functions" (McNaughton, 1975). Other investigators who have utilized the concept in one way or another in the past several years include Gadgil and Bossert (1970), Gadgil and Solbrig (1972), Barclay (1974), Forsyth and Robertson (1974), and McNaughton (1975).

The use and application of the concept of r- and K-selection in considering host-parasite relationships has been non-existent until recently. Seidenberg, et al. (1974) suggested that the nematode *Longistriata adunca*, an enteric nematode of the cotton rat, *Sigmodon hispidus*, exhibited characteristics of an r-selected species since adult males died soon after copulation thus reducing the overall population density and increasing resource availability (both nutrients and space) to egg-producing females. More recently, Jennings and Calow (1975), in an excellent paper, reviewed the relationship between fecundity and the evolution of endoparasitism. They indicated that "r-strategists, with high fecundity, can be expected to have low calorific values because their resources are channeled into production of the maximum number of progeny, while K-strategists will have high calorific values based on lipid reserves which buffer adults against possible reductions in food supply." They provide evidence which suggests that the relationship between fecundity and caloric content in r- and K-strategists can be seen after comparing various members of the phylum Platyhelminthes, some of which are parasitic and others free-living. Endoparasitic flatworms exhibit high fecundity in combination with low caloric values, while free-living species show the opposite (low fecundity and high caloric values); ectocommensalistic forms were reported to be similar to free-living platyhelminths. They argue, however, "that parasites, particularly endoparasites, follow both an r- and K-strategy at the same time and that this is only possible because of the stable, nutrient-rich environment provided by the host. Evolutionary theory dictates that all species would follow an r- and K-strategy simultaneously but environmental conditions force them into one alternative or another. Consequently the high fecundity of endoparasites, which has hitherto been viewed as a specific adaptation to endoparasitism, is now viewed as an automatic consequence of the conditions provided by the parasitic environment."

A traditional approach among parasitologists has been to view high fecundity as an adaptation for a parasitic mode of life. In an earlier paper Calow and Jennings (1974) disputed this contention by saying that "an interesting consequence of our interpretation of the relationship between the nature of the food reserve, mode of life and level of fecundity, is that parasitism, and especially entoparasitism can be regarded simply as adaptive devices which favor high fecundity." They indicate that free-living flatworms tend to have higher concentrations of stored lipid with concomitant higher calorific values and lower fecundity as compared to parasitic flatworms. In the latter group, the high fat levels are reduced and replaced by higher levels of glycogen which are relatively lower in calorific values than lipid. Previous literature had suggested that the high glycogen and low lipid levels in parasitic flatworms were the result of low oxygen concentration in the enteric environment. Calow and Jennings (1974) argue that it is an "adaptation for high fecundity, necessitated by the mode of life." They point out, however, that the high glycogen level in endosymbionts does represent a pre-adaptation to relative anaerobiasis. They further indicate that while there is a direct relationship between glycogen levels, low levels of oxygen and a modified intermediary carbohydrate metabolism among endosymbiotic flatworms, "the relationship is seen to be proximate rather than ultimate in nature."

While the ideas presented by Calow and Jennings (1974) and Jennings and Calow (1975) are intriguing, they may not be completely acceptable. As pointed out by Boddington and Mettrick (1976), there are too few data which support their contention and "of the data for the free-living forms, 50% are artificially inflated because they were for animals in the well-fed state, while the values for the parasitic forms were all field data to which no nutritional significance could be attached." The latter point could be of major significance in calculating calorific values of any parasitic helminth, flatworm or otherwise. For example, starvation of *Fasciola hepatica* for 24 hours resulted in the virtually complete depletion of parenchymal glycogen, which in turn would effectively alter the relative dry weight percentage of carbohydrate, lipid and protein (von Brand and Mercado, 1961); obviously, the calorific values would also change. Calow and Jennings (1974) state, "thus our results confirm the evolutionary trend within the Platyhelminthes, assuming that the free-living forms are primitive, has been toward a reduction in the amount of lipid stored per unit weight with a concomitant reduction in calorific values." In 1966, von Brand summarized the lipids in percent of dry weight of tissues for a wide variety of parasitic flatworms. The values ranged from a low of 1.35% (Goil, 1958) for *Gastrothylax cremenifer*, a trematode of the rumen of water buffalo, to a high of 34.6% (Goodchild and Vilar-Alvarez, 1962) for *Hymenolipis diminuta*, a tapeworm of rats. Very recently, Yusufi and Siddiqi (1976) reported that the lipid in percent of dry weight of tissues of *Gastrothylax cremenifer* was 10.5%. This latter value and the one given by Goil (1958) for the same fluke are an order of magnitude different and, while conversion to calorific values was not made, it is certain that these would also be significantly different. It is also worth noting that *Fasciola hepatica*, a bile duct trematode of sheep, has a lipid content of 13.3% (Weinland and von Brand, 1926) while the closely related *Fasciolopsis buski*, an intestinal fluke of pigs, was found to have a lipid content of 50.4% (Yusufi and Siddiqi, 1976). The variability (as noted above) in lipid, as a percentage of dry weight, could be due to (1) nutritional conditions at the time of sampling, (2) reproductive condition of the parasite, (3) inherent species-specific differences in the parasites tested, (4) age of the parasite and (5) as pointed out by Yusifi and Siddiqi (ibid), the technique used for lipid extraction. Any of these variables could have influenced the findings of Calow and Jennings (1974) and, in turn, would influence the validity of their hypothesis.

While the notions presented by Calow and Jennings (1974) should be viewed with caution, they are of exceptional value if they do nothing more than stimulate debate and challenge accepted dogma among parasitologists. Furthermore, even in view of the proper criticisms by Boddington and Mettrick (1976) concerning the absence of adequate controls in their test animals, it is quite conceivable, even likely, that a relationship between r- and K-selection, relative nutrient and calorific values, and evolutionary strategies among free-living and parasitic flatworms may exist as suggested by Jennings and Calow (1975).

The relationship between r- and K-selection, parasitism, and regulation of parasite populations can also

be viewed within the broader context developed by
Dobzhansky (1950), MacArthur and Wilson (1967) and
Pianka (1970). For example the biology of colonization, succession and the vicissitudes of ecosystem
stability and complexity in relation to parasitism and
r- and K-selection could be examined, but have not.
Thus, there are several approaches which may be taken.
The framework for the present approach is provided by
the schematic model shown in Figure 1.

There are three questions which will be considered
in the following analysis. First, and probably most
important, is it feasible to consider parasitism and
parasites in terms of r- and K-selection? Second, if
parasites can be characterized in this manner, can they
be identified as more r-selected or K-selected, or do
they constitute a spectrum between the extremes and
thus parallel the r-K continuum? And third, are there
any unique life history strategies associated with
parasites in general which might re-enforce any general
conclusions which may be made regarding the case for
parasites as being either r- or K-selected?

The aim of the analysis will not be simply to
identify selection strategies in association with particular parasite species. The objective instead will
be to examine the strategies in such a way that a different and, perhaps, more useful perspective may be
gained with regard to the behavior and regulation of
parasite populations within a variety of environmental
conditions.

An ideal point of departure for considering the
questions posed above is to analyze a set of correlates
created by Pianka (1970) for the purpose of comparing
and contrasting r- and K-strategies. The correlates
are based on such variables as climate, mortality,
survivorship, population size, competition, development
time, body size, length of life, etc.

<u>Climate</u>: r-selection - "variable and/or unpredictable:
           uncertain" (Pianka, 1970).
       K-selection - "fairly constant and/or pre-
           dictable: more certain" (Pianka, ibid).

Within the biosphere there is a continuum of

environments ranging from those which are more or less constant and predictable to those which are variable and unpredictable. Organisms existing in the latter may be subjected to heavy mortality, on an irregular basis. In general, such species survive under conditions which would be less competitive than for those existing in more stable environments. The ideal strategy (r-strategy) under less stable conditions would be to produce as many progeny as possible. The more stable environment, on the other hand, would be saturated. The outcome would be intense competition with maximum energy directed at development of maintenance and competitive skills with less effort directed at production of large numbers of progeny. These species would thus exhibit characteristics of a K-strategist.

After examination of the correlate for climate, it was felt that Pianka (1970) intended to refer to the role of temperature, humidity, rainfall, soil conditions, etc., in influencing the life history and physiological characteristics of free-living organisms. For many parasitic animals, the impact of environmental variability would be similar to those of free-living organisms since most parasites have free-living life cycle stages. However, most of the life of a parasite is spent within a host and could, therefore, because of various homeostatic processes associated with the host, be considered as life within a relatively stable environment. In spite of this, it is generally true that environmental instability is the rule.

Take, for example, the life cycle of a digenetic trematode, beginning with the release of the egg into an aquatic environment (assuming, of course, that the egg requires the interposition of an aquatic environment, and that the egg is shed therein - thus the initial vagary). On hatching, a free-living larval stage (the miracidium) emerges. The miracidium is subjected to the same set of environmental variables with which the free-living organisms must also contend. These would, of course, include temperature, osmotic pressure, pH, etc. While the larval trematode has evolved a remarkable array of physiological and behavioral adaptations in order to survive such conditions, it is unquestionably true that the majority die in a relatively brief period of time (see section on survivability).

Utilizing chemotatic responses, the miracidium locates and penetrates an appropriate intermediate host and therein is subjected to significant environmental changes, including especially pH and osmotic pressure. Perhaps even more significant, the parasite is subjected to the potential of a host response which could result in mortality. In some cases, there is even evidence for predation of newly penetrated trematodes by previously established larval stages of other digenetic trematodes (Lie, Heyneman, and Kostanian, 1975). If the miracidium is successful in becoming established, dramatic morphological and physiological changes then occur with the outcome being the development of asexually reproducing sporocysts and/or redia, depending upon the species. The intramolluscan larvae then produce yet another larval stage which exists from the molluscan sanctuary. These larvae, called cercaria, enter the aquatic environment and, like miracidia, must cope with significant changes in temperature, pH and osmotic pressure. And, as with miracidia, cercaria survivability is low and longevity is brief (on the order of a few hours).

Employing a complex set of physiological and morphological adaptations, the cercaria seek out and penetrate an appropriate intermediate host. (Note that this route varies according to parasite species since some cercaria may encyst in the open on vegetation or other substrates, while other cercaria may penetrate directly the definitive host). On penetration of the intermediate host, the parasite usually encysts, becoming a metacercaria. Within the host the parasite is subject, once again, to significant changes in temperature, pH and osmotic pressure. And, again, the parasite must cope with host immune responses. The metacercaria then remains until the intermediate host is consumed by the appropriate definitive host. While the metacercaria is essentially quiescent, there are indications that the host continues to respond to the parasite and that, at least in some cases, the parasite may be overwhelmed and killed in the process.

When finally ingested by the definitive host, the metacercaria is subjected to a new set of environmental changes which, in many ways, may be the most hostile. Thus, when the metacercaria excysts in the stomach, it encounters a pH of about 1-3 and a complex of protein-digesting enzymes. On entering the small intestine, the parasite is exposed to new environmental conditions, with different enzymes, an array of bile salts and a pH of 7.0. If the parasite and host are suitable to each other, the trematode will undergo sexual maturation and egg production will follow quickly.

From the foregoing account, it should be clear that the parasite must be well adapted to the rapid and dramatic shifts in environmental conditions which occur throughout the life cycle. This is also obvious from the literature which shows the large number of parasites in animals from virtually all possible habitats (Yamaguti, 1958). This description of a generalized, trematode life cycle, however, also serves as an indication of the high degree of variability and unpredictability of the parasite's environment. It is safe to state that most parasite species are exposed to more environmental or climatic variability than virtually any other group of animals. From the standpoint of climate, therefore, it must be concluded that parasites are subject to selection forces which are characteristic of r-strategists.

<u>Mortality</u>: r-selection - "often catastrophic, non-directed, density-independent" (Pianka, 1970).
K-selection - "more directed, density-dependent" (Pianka, ibid).

At both the suprapopulation and infrapopulation levels, parasites are regulated by density-dependent and density-independent processes. For evidence of density-independent regulation of parasite infrapopulations, the study by Kennedy (1974) on the acanthocephalan *Pomphorhynchus laevis* is illuminating. Utilizing the common goldfish as an experimental host, Kennedy reported that when "infections with different population densities of parasite were given to fish on one occasion only and on several occasions with different time intervals between infection," that "the parasites established and grew in experimental hosts at rates comparable to those in natural hosts on all occasions." He stated further that "the proportion establishing did not increase with increasing population size and bore no relationship to the presence of an existing population." He concluded that parasite mortality was not subject to feedback control and was of little consequence in regulating the size of parasite infrapopulations. Awachie (1966) reached similar conclusions regarding the regulation of *Echinorhynchus truttae* densities in *Salmo trutta*.

Recognizing the potential for both positive and negative relationships between hosts and parasites, Bradley (1974) proposed three mechanisms by which parasite populations are regulated. Bradley's Type I regulation involved factors which are related to transmission of the parasite from one host to another; essentially, this form of regulation would be density-independent and would be controlled by extrinsic, environmental forces. An example given by Bradley was of "malaria in an epidemic area, such as the Punjab, where ambient temperature, humidity, distribution of surface waters, vegetation shading those waters, number insectivorous birds, may all affect transmission but are scarcely or not at all affected by transmission."

Those situations in which parasites are regulated at the host population level are referred to as Type II regulation (Bradley, 1974). In a very real sense, this kind of regulation is density-dependent and is related to two phenomena, i.e. immunity and overdispersion. Since overdispersion will be considered subsequently, it will be only briefly alluded to here. According to Crofton (1971), the overdispersion of parasite infrapopulations is an inherent property of host-parasite systems. In effect, most parasites within a group of infrapopulations are found within a small number of hosts. Because of overdispersion, there is a potential to increase the density of a given infrapopulation to such an extent that it will eventually cause the host to die. If this occurs, then the density of the infrapopulations is reduced. Hence, there is negative feedback by the parasite; overdispersion (and, therefore, density) is the operational factor. Bradley (1974) did suggest, however, that overdispersion may not produce stability for densities of helminth parasites among vertebrate hosts though it may for parasites among invertebrate populations. Immunity (in this case, complete immunity) was also considered as a density-dependent factor in regulating parasite populations. Thus, a situation could arise in which a single infrapopulation would become large enough to induce complete immunity in a host and preclude subsequent reinfection. Should this occur, the impact would be the same to the parasite as if the host were non-existent, or dead. In effect, the density of the parasite infrapopulations would be self-regulated except that the host population would be mediating the regulation through the immune response.

Type III regulation is considered by Bradley (1974) as the most efficient in maintaining stability of parasite infrapopulations since the densities are regulated by individual hosts. Thus, he states, "consider a helminth whose usual transmission rate is 100 worms inoculated per host and the host has a means of preventing the worm load exceeding 10, then over a tenfold fall or indefinite rise in transmission the parasite population will be perfectly regulated. This is the ultimate in regulation: highly efficient transmission combined with 'premunition' or some similar process of parasite regulation by each host." This form of regulation would be density-dependent and, according to Bradley, "is seen to a varying degree in many parasitic systems."

It is apparent that both regulatory mechanisms (density-dependent and density-independent) operate, depending on the specific host-parasite system involved. Indeed, it is conceivable that both may operate for a given host-parasite system, with one form of regulation influencing the density of parasites within an intermediate host and the other form at the level of the definitive host. It is probably true, that with respect to this correlate, parasitic organisms lie along the r-K continuum and not exclusively at either endpoint.

<u>Survivorship</u>: r-selection - "Often Type III (Deevey, 1947)," Pianka (1970).
K-selection - "Usually Type I and II (Deevey, 1947)," (Pianka, ibid).

Type I survivorship (Pearl, 1928) is reflected by an age specific death rate which is expressed by a hyperbolic function. Western man exhibits a Type I survivorship curve, i.e. there is relatively low mortality in early age classes with subsequent mortality continuing to remain low until near the limit of normal longevity. The Type II survivorship curve is linear (Caughley, 1966), with mortality rates proceeding at the same rate regardless of age; birds commonly exhibit such a survivorship curve. According to Pianka (1970), these two types of survivorship are characteristic of K-selected species.

A Type III survivorship curve follows an inverse hyperbolic function; it is produced by high initial mortality with a gradual decline and subsequent lower mortality among older individuals. According to Caughley (1966), this kind of survivorship is typical of fish and insects. Type III survivorship is characteristic of r-selected species, although Pianka (1970) also states that fish may span the entire r-K continuum.

Intuitively, it might be guessed that survivorship of parasites would be best characterized by a Type III curve. However, there have been too few studies on age specific mortality throughout an entire life cycle. Such an assessment is further complicated since many parasite species have life cycle stages in which asexual reproduction occurs. Anderson and Whitfield (1975) noted a high initial mortality of cercaria of *Transversotrema patialensis*, a trematode using fish as the definitive host. After 26 hours, survivorship was about 50%; this was followed by complete mortality after 44 hours as glycogen reserves were depleted. In this case, survivorship was not linear and might be best considered as a modified, Type III curve.

At the suprapopulation level, it is speculated that survivorship for most parasite species probably fits a Type III curve. Intuitively, this would be correct in view of the prodigious reproductive capacities of most parasites as compared with the relatively low chances for success of a single egg completing the entire cycle. In view of the survivorship criterion, it seems appropriate to characterize most parasitic organisms as r-strategists.

Population size: r-selection - "Variable in time, nonequilibrium; usually well below the carrying capacity of the environment; unsaturated communities or portions thereof; ecologic vacuums; recolonization each year." (Pianka, 1970).

K-selection - "Fairly constant in time; at or near carrying capacity of the environment; saturated communities, no recolonization necessary." (Pianka, ibid).

Populations of r-strategists are generally of variable nature, at well below the carrying capacity. Typically, r-strategists are capable of rapid, annual recolonization which also implies they will move into an ecological vacuum with facility. In terms of host-parasite systems, it is reasonable to consider a non-parasitized host as an ecological vacuum and, for many parasite species, recolonization is a typical annual strategy. Many helminth parasites are seasonal, infecting the definitive host in spring, producing eggs, and senescing in late summer or fall. The densities of many parasite infrapopulations are thus highly variable within a single 12 month period.

While annual recolonization is a typical strategy for many parasite species, this tendency may well be tempered by the state, or level, of existing immunity within the potential host. As suggested by Bradley (1974), the level of immunity, or 'premunition', may be one of the most significant elements in regulating parasite densities. Indeed, the level of immunity, in conjunction with finite space and nutrient resources, may be considered as the primary limiting factor in determining the carrying capacity of an individual host. In many cases, it is conceivable that space and nutrient resources would not ever be fully exploited because of the overriding effect of immunologic factors. Under these conditions, the density of the parasite infrapopulation would remain below the 'apparent' carrying capacity of the host. It is also possible that some parasite infrapopulations may never reach the 'apparent' carrying capacity, since death of the host may occur before K is reached. On the other hand, as has already been suggested, some parasite populations remain in a state of dynamic equilibrium on an annual basis, i.e. recruitment and turnover within the parasite infrapopulations are constant.

Since very little is known about carrying capacities for parasite infrapopulations under natural conditions, this correlate must be viewed with some caution. It would appear, however, that most parasite species exhibit characteristics of an r-strategist when such factors as density variability, equilibrium versus non-equilibrium, recolonization, etc., are considered.

<u>Intra- and interspecific competition:</u> r-strategist - "Variable, often lax" (Pianka, 1970).
K-strategist - "Usually keen" (Pianka, ibid).

Both intra- and interspecific competition are known to occur among and between various parasite species. Roberts (1966) reported stunting of *Hymenolepis diminuta* when infrapopulations were at high densities. Ghayal and Avery (1974) indicated that when infrapopulation densities were high, that length, weight and egg production decreased for cysticercoid-derived and egg-derived infections of the cestode *Hymenolepis nana*. The tapeworm *Diphyllobothrium dendriticum* is smaller when at high infrapopulation densities in both hamsters and gulls (Halverson and Anderson, 1974). Holmes (1961) showed that densities as low as 10 worms per host would result in both reduced length and weight of *Hymenolepis diminuta*.

In an elegant study, Holmes (1961) conclusively demonstrated competitive exclusion of one parasite species by another. In concurrent infections by the tapeworm *Hymenolepis diminuta* and the acanthocephalan *Moniliformis dubius* in the white rat, he was able to show clearly that (1) growth of the tapeworm was retarded and (2) the tapeworm attached to the wall of the intestine at a site other than the one observed in single species infections. He also showed that established populations of *H. diminuta* migrated away from the initial site of infection on the appearance of *M. dubius* at that site. While Holmes' study was conducted under laboratory conditions, Chappel (1969) provided field evidence for competitive exclusion of the tapeworm *Proteocephalus filicollis* by the acanthocephalan *Neoechinorhynchus rutili* in three-spined sticklebacks. In both of these studies, competitive interactions resulted in longitudinal redistribution of the affected parasite species. Working with the European tortoise, *Testudo graeca*, Schad (1963) clearly demonstrated that competitive interaction would also effect changes in the radial distribution of several species of enteric, parasitic nematodes and that the apparent operational factor in producing the changes in location was likely due to differential food preferences among the affected species.

From the studies cited above, it is apparent that intra- and interspecific competition do occur among various species of parasitic helminths. Indeed, Kennedy (1975) states that "nearly all species of cestodes appear to show crowding effects." In spite of this sweeping statement, just how common competition is among different species of parasitic organisms is actually difficult to assess, since far too few investigators have reported on efforts to obtain evidence which would address this question. Since overdispersion is an inherent characteristic of parasitism and, thus, the potential for competitive interaction is certainly extant for most parasitic organisms, there is opportunity for generating data to provide support for a generalization such as Kennedy's (ibid). Until more information is available, it would seem judicious to wait before attempting to categorize parasites as r- or K-selected in terms of intra- and interspecific competition.

Longevity: r-strategists - "Short, usually less than
             one year" (Pianka, 1970).

K-strategists - "Longer, usually more than one year" (Pianka, ibid).

When the correlate for population size was being discussed, it was stated that seasonal fluctuation in population density was a common attribute for parasites. In part, this observation is correct because the generation time of most parasites is less than one year. There is, however, some variability. For example, the trematode *Fascioloides magna* may survive for more than five years in the livers of deer (Blazek, Erhadová-Koctila, and Kortley, 1972), while the closely related fluke *Fasciola hepatica* has a life span of only six months in sheep (Boray, 1969). Life spans of the nematodes *Ascaridia dissimilis*, *Heterakis gallinarum*, *Capillaria caudinflata* and *C. obsignata* are all less than nine months while that of *Syngamus trachea* was shown to be 126 days in turkeys and 92 days in chickens (Barus, 1966). Other parasites, such as some taeniids and schistosomes, are said to be rather long-lived, spanning several years.

It is safe to say that the life span of most helminth parasites is less than one year; the same is true for most protozoa which may have a life span ranging from a few hours to a few days. It would thus be reasonable to conclude that most parasites exhibit an r-strategy in terms of life span.

OTHER CONSIDERATIONS

At the outset of the discussion of r- and K-selection and parasitism, three questions were posed. The first was directed at the feasibility of fitting parasites within the concept of r- and K-selection. The second question referred to the position of parasites within the r-K continuum. Based on the foregoing analysis, it is, first, clear that parasites can be considered within the conceptual framework provided by the notions of r- and K-selection and, second, it is safe to conclude that selection among parasite species has favored the evolution of life history strategies, morphological characteristics and physiological traits which *collectively* suggest the spector of r-selection.

This statement must be qualified by noting that some of these features suggest that certain parasites are not at the r- endpoint, but instead lie along the r-K continuum.

The third question was directed at the possible existence of any unique attributes of parasites which would re-enforce conclusions regarding the nature of selection forces which may affect parasites. There are several attributes which may be considered as unique. Foremost among these is the phenomenon of dispersion.

The dispersal and dispersion patterns of a population, or the description of how an organism is dispersed in space and time and the pattern of the eventual distribution of these organisms, are of considerable ecological importance (Cassie, 1962). In describing the dispersion of organisms within a particular habitat, there are three general types of frequency distribution which are recognized. These are (1) a uniform, or underdispersed, population, (2) a randomly distributed population, or (3) an aggregated, or overdispersed, population. On examination of the dispersion patterns of various populations, it has become increasingly evident that most species do not exhibit a uniform, or underdispersed distribution. Exceptions to this generalization might arise when competition for resources is severe or when chemical substances might be produced such that other organisms are prevented from becoming established nearby. If a population is randomly distributed, then the mean density and variance, per unit area, will be equal. Random distributions can be described by the Poisson frequency distribution model. Randomly distributed populations are also apparently rare in nature (Cassie, 1962). The most commonly observed frequency distribution pattern is one of overdispersion, i.e. contagion. In this type of distribution, organisms within the population are found in groups or clumps and the variance exceeds the mean density.

Even though the vast majority of studies on population distributions have focused on free-living organisms, dispersion analysis is also applicable to parasites. While Li and Hsu (1951) were among the first to examine the frequency distributions of parasites within a host population, Crofton (1971) was the first to employ parasite dispersion as a means for explaining regulation of parasite populations. Through his efforts and those of more recent investigators (James and Srivastavy, 1967; Pennycuick, 1971; Schmid and Robinson, 1972; Boxshall, 1974; Anderson, 1974), it has been clearly shown that overdispersion is an inherent characteristic of parasite systems.

A number of theoretical models, e.g. Polya-Aeppli (Polya, 1931), double Poisson (Thomas, 1949), Neyman Type A (Neyman, 1939), log series, log normal, and negative binomial, have been used to describe the frequency distributions of parasite infrapopulations. However, in most cases where studied, the negative binomial model has provided the best fit to the observed distribution.

According to Pennycuick (1971) (also see Crofton, 1971), the distribution of parasite infrapopulations within a host population "may be a negative binomial if there is a departure from randomness caused by any of the following factors:
1. The host is exposed to several waves of infection, each of which attacks randomly, giving rise to a series of Poissons.
2. The infective stages of the parasite are not randomly distributed.
3. The presence of a parasite in a host increases or decreases its chances of acquiring further infections.
4. The sampling units are unequal; for example, the hosts are of different ages.
5. The sampling units change during sampling; for example, if the sampling takes a long time, the age of the hosts will change."

The concept of overdispersion has proved quite useful in understanding the regulation of parasite populations (Crofton, 1971). It is also possible to view the overdispersion concept in terms of competition and, thus, r- and K-selection. Pianka (1970) stated that selection for a K-strategist favors slower development, lower resource thresholds, delayed reproduction, larger body size and iteroparity. These characteristics maximize competitive ability. Conversely, selection for an r-strategist favors rapid development, higher resource thresholds, rapid reproduction, smaller body size and semelparity which, collectively, minimize competitive ability. Most parasites exhibit each of these latter characteristics. Thus, they develop rapidly in response to appropriate environmental cues, they generally reproduce quickly after reaching sexual maturity, they are small in body size and many are semelparous, that is to say, they reproduce but once in their lifetime. Even tapeworms, which are usually considered as being iteroparous, or continuous reproducers, may be thought of as semelparous since sexual reproduction in each proglottid apparently occurs only once.

It can be argued that overdispersion of parasite infrapopulations results in a reduction, or minimization, of competition. At first glance, this assertion may appear paradoxical, especially when it is noted that in many cases, 50-60% of the parasites in a given host population may be recovered from as few as 20-30% of the hosts. For example, approximately 60% of the metacercaria of *Crepidostomum cooperi* are found in 40% of the subimagoes of the burrowing mayfly, *Hexagenia limbata*, at the time of their emergence from Gull Lake, Kalamazoo County, Michigan (Esch, et al., unpublished observations). In this case, the metacercaria are overdispersed and the distribution can be fitted to the negative binomial (the mean metacercaria densities are 7.48 per female mayfly and the variance is 54.76; N = 1100). It must be noted, however, that in this, as well as other host-parasite systems, most of the hosts (70-80%) are fully exploitable in the sense that both space and nutrient resources are capable of sustaining recruitment of new parasites as well as maintaining established infrapopulations. It is also possible, if not probable, that this 70-80% of the hosts are the most likely to be immunologically ineffective in preventing further recruitment of parasites into the infrapopulation. While the majority of parasites would be found in hosts where resources could become limiting, the high r_{max} of *all* parasites, including especially those within the relatively unexploited 70-80% of the host population, would be more than likely capable of supplying the reproductive requirements of the suprapopulation, even if overdispersion produces death in heavily parasitized hosts. Thus, the advantage of overdispersion to the parasite would be in restricting, or reducing, competition in a relatively few infrapopulations. The advantage of such a strategy is clear, since most of the space and nutrient resources required by the parasite species would remain available for exploitation to a substantial portion of the parasites. If overdispersion can be considered in these terms, then it is concluded that it minimizes competition and, in doing so, is a strategy characteristic of r-selection.

Many parasite species display other life history characteristics which may also be considered as strategies for avoiding or minimizing competition so that maximum utilization of resources can be effected. One of the more unique ploys is associated with the host immune responsiveness. It would generally be agreed that immune responsiveness of a host to a parasite is a useful mechanism to protect against over-infection. While probably evolved as a defense system by the host, the outcome of immunity may also provide survival value for the parasite. Bradley (1974) has already pointed out that the host immune response should be considered as an important mechanism in regulation of parasite populations and, thus, in producing stability. Since the immune response is also effective in preventing heavy parasite loads within a host, the resulting reduction in size of the infrapopulation also minimizes the potential for competitive interaction among parasites and should, therefore, be considered as an r-strategy by parasitic organisms (mediated, of course, by the host).

There are a number of mechanisms which insure that parasitic animals have mating partners. These include hermaphroditism, monoeciousness, and protandrous hermaphroditism. For parasitic animals employing such reproductive strategies, there is survival value. An obvious advantage is that mating partners are guaranteed. However, the same strategies also minimize the competition for resources. Thus, the density requirement of a breeding infrapopulation could be reduced by as much as half for species of parasites which reproduce in one of the ways listed above. The same infrapopulation would also cut by a similar fraction the requirement for space and nutrient resources. Disproportionate sex ratios and early death of male parasites following copulation (Chan, 1952; Crofton and Whitlock, 1969; Seidenberg, et al., 1974) could also be considered as selection mechanisms which would minimize the potential for competitive interaction among some parasite species.

The general conclusion has been that parasitic animals exhibit many characteristics of an r-strategist. It is also clear that some species cannot be categorized more toward one end-point than the other and, therefore, occupy a position along the continuum. Indeed, it also seems conceivable that within a single suprapopulation, there could be a variety of genotypes, representing a genome with a high degree of variability with respect to r- and K-strategies. Genotypic variability would be of interest if parasitism is examined in terms of stress, or in terms of epizootic outbreaks within a given ecosystem.

Though somewhat afield of parasitism, Gadgil and Solbrig (1972) presented quantitative evidence that a population of one species of dandelion, occupying a given space, was actually separated into several ecotypes, differing in seed output, proportion of biomass devoted to reproductive tissues, and competitive ability, according to the degree of perturbation within the area. In effect, some ecotypes were more typical of r-strategists, while others were more characteristic of K-strategists.

If the observations of Gadgil and Solbrig (ibid) can be extended to host-parasite systems, then it is possible to see how the concept of r- and K-selection could be used in evaluating the relative stability of a parasite suprapopulation within an ecosystem, or in how the concept could be employed in explaining sudden and unpredicted epizootics. For example, it is conceivable that r- and relatively K-selected biotypes of a parasite species might co-exist within the same suprapopulation. Under more or less constant environmental conditions, individuals exhibiting a predominantly K-strategy would be selected for and the parasite suprapopulation would remain stable. While the K-strategists would dominate within the ecosystem, some individuals, possessing characteristics of an r-strategists would nonetheless persist. Then, suppose the ecosystem is perturbed in a manner in which, rapidly, selection favors the allocation of resources into more progeny rather than fewer. Under such circumstances, the r-strategist would be favored. The result would be a sudden, exponential expansion of the population and, in effect, an epizootic outbreak. In cases where epizootics have been known to occur, a sudden crash in the density of the parasite population has also been reported. In some instances, the sudden decline in parasite density could be associated with the withdrawal or removal of the perturbation. In these situations, selection forces might again favor K-strategists within the suprapopulation. Ultimately, the suprapopulation density would stabilize on reaching the K-asymptote, or carrying capacity. While selection would continue to favor the K-strategist, some r-strategists would remain and the entire sequence of events could reoccur should conditions change.

CONCLUDING REMARKS

To summarize, the basic aim of this presentation has not been to develop a case for support or rejection of the notions or concepts associated with r- and K-selection. Rather, the aim was to examine the physiologic, morphologic and life history characteristics of a functionally similar group of animals, namely parasites, and determine if the analysis will permit a comparison of this group with other groups which may appear dissimilar but which may nonetheless exhibit similar selection strategies. Perhaps such an approach will contribute to the development, or modification, of current views of regulation as they pertain to parasite population biology.

ACKNOWLEDGEMENTS

We want to express our appreciation to Mrs. Lila Spencer and Mrs. Yancey Smith for their tireless effort in deciphering and typing the manuscript. We also thank Mrs. Jean Coleman for preparation of the figures. Preparation of this manuscript was supported, in part, by Contract E(38-1)-900 between Wake Forest University and the United States Energy Research and Development Administration (ERDA) and by Contract E(38-1)-819 between the University of Georgia and ERDA. We should also like to thank Dr. Ronald Dimock for his reading and criticism of the manuscript.

LITERATURE CITED

ADDIS, J. C. 1946. Experiments on the relations between sex hormones and the growth of tapeworms *(Hymenolepis diminuta)* in rats. *J. Parasitol. 32:* 574-580.

AHO, J. M., J. W. GIBBONS, and G. W. ESCH. In press. Relationship between thermal loading and parasitism in the mosquitofish, *Gambusia affinis. In:* G. W. Esch and R. W. McFarlane (eds.), Thermal Ecology-II. ERDA Symposium Series (CONF. 750425).

ANDERSON, R. M. 1974. Population dynamics of the cestode *Caryophyllaeus laticeps* in the bream. *J. Anim. Ecol. 43(2):* 305-321.

ANDERSON, R. M. and P. J. WHITFIELD. 1975. Survival characteristics of the free-living cercarial population of the ectoparasitic digenean *Transversotrema patialeusis* (Soparker, 1924). *Parasitology 70:* 235-310.

ANDREWARTHA, H. G. 1971. Introduction to the Study of Animal Populations. University of Chicago Press, Chicago.

ARME, C. and R. W. OWEN. 1967. Infections of the three-spined stickleback, *Gasterosteus aculeatus* L., with the plerocercoid larva of *Schistocephalus solidus* (Muller, 1776), with special reference to pathological effects. *Parasitology 57:* 301.

AWACHIE, J. B. E. 1966. The development and life history of *Echinorhynchus truttae* Schrank, 1788 (Acanthocephala). *J. Helminthol. 40:* 11-32.

BARCLAY, H. 1975. Population strategies and random environments. *Can. J. Zool. 53:* 160-165.

BARUS, V. 1966. The longevity of the parasitic stages and the dynamics of egg production of the nematode *Syngamus trachea* (Montagu, 1911) in chicken and turkeys. *Folia Parasitol. 13:* 274-277.

BAUER, O. W. 1959. The influence of environmental factors on reproduction of fish parasites. From: *Vopr. Ecol.* (Izdatelstro Kierskoyo Universiteta), *3:* 132-141.

BECK, J. W. 1952. Effect of gonadectomy and gonadal hormones on singly established *Hymenolepis diminuta* in rats. *Exp. Parasitol. 1:* 109-117.

BLAZEK, K., B. ERHARDOVA-KOTILA, and A. KORTLY. 1972. Ovulation of the trematode *Fascioloides magna* in relation to the duration of parasitism. *Folia Parasitol. 19:* 335-339.

BODDINGTON, J. F. and D. R. METTRICK. 1976. Seasonal changes in the biochemical composition and nutritional state of the immigrant triclad *Dugesia polychroa* (Platyhelminthes:Turbellaria) in Toronto Harbor, Canada. *Can. J. Zool. 53:* 1723-1734.

BORAY, J. C. 1969. Experimental fascioliasis in Australia. *Adv. Parasitol. 7:* 96-210.

BOURQUE, J. E. and G. W. ESCH. 1974. Population ecology of parasites in turtles from thermally altered and natural aquatic communities. *In:* J. W. Gibbons and R. R. Sharitz (eds.), Thermal Ecology. AEC Symposium Series (CONF-730505).

BOXSHALL, G. A. 1974. The population dynamics of *Lepeophtheirus pectoralis* (Muller): dispersion pattern. *Parasitology 69:* 373-390.

BRADLEY, D. J. 1974. Stability in host-parasite systems. *In:* M. B. Usher and M. H. Williamson (eds.), Ecological Stability. Chapman and Hall, London.

CALOW, P. and J. B. JENNINGS. 1974. Calorific values in the phylum platyhelminthes: the relationship between potential energy, made of life and the evolution of entoparasitism. *Biol. Bull. 147:* 81-94.

CANNON, L. R. G. 1972. Studies on the ecology of the papillose allocreadid trematodes of the yellow perch in Algonquin Park, Ontario. *Can. J. Zool. 50:* 1231-1239.

CASSIE, R. M. 1962. Frequency distribution models in the ecology of plankton and other organisms. *J. Anim. Ecol. 31:* 65-92.

CAUGHLEY, G. 1977. Mortality patterns in mammals. *Ecology 47:* 906-918.

CHAN, K. F. 1952. Life cycle studies on the nematode *Syphacia obveleta*. *Amer. J. Hyg. 56:* 14-21.
CHAPPEL, L. H. 1969. The parasites of the three-spined stickleback *Gasterosteus aculeatus* L. from a Yorkshire pond. I. Seasonal variation of parasite fauna. *J. Fish Biol. 1:* 137-152.
CHUBB, J. C. 1967. A review of seasonal occurrence and maturation of tapeworms in British freshwater fish. *Parasitology 53:* 13.
CLARKE, H. V. de V. 1965. The relationship between acquired resistance and transmission of *Schistosoma* Weinland, 1858, in man, and its influence on the prevalence of *S. capense* (Hanley, 1864) and *S. mansoni* Sambon, 1907, in Southern Rhodesia. Ph.D. Thesis, Rhodes University, South Africa.
CLARKE, K. R. 1968. Effect of low protein diet and a glucose and filter paper diet on the course of infection of *Nippostrongylus brasiliensis*. *Parasitology 58:* 325-339.
CONNOR, R. S. 1953. A study of the seasonal cycle of a proteocephalan cestode, *Proteocephalus stizostethi* Hunter and Baugham, found in the yellow pike perch, *Stixostedion vitreum vitreum* (Mitchill). *J. Parasitol. 39:*621-624.
CROFTON, H. D. 1971. A model of host-parasite relationships. *Parasitology 63:* 343-364.
CROFTON, H. D. and J. H. WHITLOCK. 1969. Changes in the sex ratios in *Haemonchus contortus cayugensis*. *Cornell Vet. 59:* 388-392.
CROLL, N. A. 1975. Behavioral analysis of nematode movement. *Adv. Parasitol. 13:* 71-122.
CULBRETH, K. L., G. W. ESCH, and R. E. KUHN. 1972. Growth and development of larval *Taenia crassiceps* (Cestoda). III. The relationship between larval biomass and uptake and incorporation of ^{14}C-leucine. *Exp. Parasitol. 32:* 272-281.
DAMIAN, R. T. 1964. Molecular mimicry: antigen sharing by parasite and host and its consequences. *Amer. Nat. 98:* 129-149.
DEEVEY, E. S. 1947. Life tables for natural populations of animals. *Quart. Rev. Biol. 22:* 283-314.
DOBBEN, W. H. VAN. 1952. The food of the cormorant in the Netherlands. *Ardea 40:* 1-63.

DOBZHANSKY, T. 1950. Evolution in the tropics. *Amer. Sci. 38:* 208-221.

DOGIEL, V. A. 1964. General Parasitology. (English translation by Z. Kabata). Oliver and Boyd, Edinburgh and London.

DONGES, J. 1963. Reizphysiologische Untersuchungen and der Cercarie von *Posthodiplostomum cuticola* (V. Nordmann 1832) Dubois 1936, dem Erreger des Diplostomatiden-Melanoms der Fische. *Verh. dt. zool. Ges.* 216-223.

ELTON, C. S. 1927. Animal Ecology. Macmillan, New York.

ESCH, G. W. 1967. Some effects of cortisone and sex on the biology of coenuriasis in laboratory mice and jackrabbits. *Parasitology 57:* 175-179.

ESCH, G. W. 1971. Impact ecological succession on the parasite fauna in centrarchids from oligotrophic and eutrophic ecosystems. *Amer. Mid. Nat. 86:* 160-168.

ESCH, G. W., G. C. CAMPBELL, R. E. CONNERS, and J. R. COGGINS. In press. Recruitment of helminth parasites by bluegill sunfish (*Lepomis macrochirus*) using a modified live-box technique. *Trans. Amer. Fish. Soc.*

ESCH, G. W. and J. W. GIBBONS. 1967. Seasonal incidence of parasitism in the painted turtle, *Chrysemys picta*. *J. Parasitol. 53:* 818-821.

ESCH, G. W., J. W. GIBBONS, and J. E. BOURQUE. 1975. An analysis of the relationship between stress and parasitism. *Amer. Mid. Nat. 93(2):* 339-353.

ESCH, G. W., T. C. HAZEN, R. V. DIMOCK, and J. W. GIBBONS. In press. Thermal effluent and the epizootiology of *Epistylis* (Peritricha:Ciliata) and the gram-negative bacterium, *Aeromonas hydrophila* in association with centrarchid fish. *Trans. Amer. Micros. Soc.*

ESCH, G. W., W. C. JOHNSON, and J. R. COGGINS. 1975. Studies on the population biology of *Proteocephalus ambloplitis* (Cestoda) in the smallmouth bass. *Proc. Okla. Acad. Sci. 55:* 122-127.

EURE, H. E. and G. W. ESCH. 1974. Effects of thermal effluent on the population dynamics of helminths in largemouth bass. *In:* J. W. Gibbons and R. R.

Sharitz (eds.), Thermal Ecology. AEC Symposium Series (CONF-730505).

FISCHER, H. 1967. The life cycle of *Proteocephalus fluviatilis* Bangham (Cestoda) from smallmouth bass, *Micropterus dolomieui* Lacepede. *Can. J. Zool. 46:* 569-579.

FISCHER, H. and R. S. FREEMAN. 1969. Penetration of parenteral plerocercoid of *Proteocephalus ambloplitis* (Leidy) into the gut of smallmouth bass. *J. Parasitol. 55:* 766-774.

FORSYTH, A. B. and R. J. ROBERTSON. 1974. K reproductive strategy and larval behavior of the pitcher plant sarcophagid fly, *Blaesoxipha fletcheri*. *Can. J. Zool. 53:* 174-179.

GADGIL, M. and W. H. BOSSERT. 1970. Life historical consequences of natural selection. *Amer. Nat. 104:* 1-24.

GADGIL, M. and O. T. SOLBRIG. 1972. The concept of r- and K-selection: evidence from wild flowers and some theoretical considerations. *Amer. Nat. 106:* 14-31.

GERKING, S. D. 1962. Production and food utilization of bluegill sunfish. *Ecol. Monogr. 32:* 50-60.

GHAYAL, A. M. and R. A. AVERY. 1974. Population dynamics of *Hymenolepis nana* in mice: fecundity and the crowding effect. *Parasitology 69:* 403-415.

GOIL, M. M. 1958. Fat metabolism in trematode parasites. *Z. Parasitkde. 18:* 321-323.

GOODCHILD, C. G. and C. M. VILAR-ALVAREZ. 1962. *Hymenolepis diminuta* in surgically altered hosts. II. Physical and chemical changes in tapeworms grown in shortened small intestines. *J. Parasitol. 48:* 379-383.

HAIRSTON, N. G. 1965. On the mathematical analysis of schistosome populations. *Bull. World Health Org. 33:* 45-62.

HAIRSTON, N. G., D. W. TINKLE, and H. M. WILBUR. 1970. Natural selection and the parameters of population growth. *J. Wildl. Manag. 34(4):* 681-698.

HALVERSON, O. and K. ANDERSON. 1974. Some effects of population density in infections of *Diphyllobothrium dentriticum* (Nitzsch) in the golden hamster (*Mesocricetus auratus* Waterhouse) and the

common gull *Larus canus* L. *Parasitology* 69: 149-160.
HARPER, J. L. 1967. A Darwinian approach to plant ecology. *J. Ecol.* 55: 247-270.
HINE, P. M. and C. R. KENNEDY. 1974. The population biology of the acanthocephalan *Pomphorhynchus laevis* (Muller) in the River Avon. *J. Fish. Biol.* 6: 665-679.
HOLMES, J. C. 1961. Effects of concurrent infections on *Hymenolepis diminuta* (Cestoda) and *Moniliformis dubius* (Acanthocephala). I. General effects and comparison with crowding. *J. Parasitol.* 47: 209-216.
HOLMES, J. C. and W. M. BETHEL. 1972. Modification of intermediate host behavior by parasites. *In:* E. U. Canning and C. A. Wright (eds.), Behavioral Aspects of Parasite Transmission. *J. Limn. Soc.* 51: Suppl. I: 123-149.
HORAK, I. G. 1971. Paramphistomiasis of domestic ruminants. *Adv. Parasitol.* 9: 33-72.
JAMES, B. L. and L. P. SRIVASTAVA. 1967. The occurrence of *Podocotyle atomon* (Rud., 1802) (Digenea), *Bothriocephalus scorpii* (Müller, 1776) (Cestoda), *Contracecum elavatum* (Rud., 1809) Nematoda and *Echinorhynchus gadi* Zoega, in Müller, 1776 (Acanthocephala) in the five-bearded rocking, *Onos mustela* (L.). *J. Nat. Hist.* 1: 363-372.
JENNINGS, J. B. and P. CALOW. 1975. The relationship between high fecundity and the evolution of entoparasitism. *Oecologia* 21: 109-115.
KAGAN, I. G. 1966. Mechanisms of immunity in trematode infection. *In:* E. J. L. Soulsby (ed.), Biology of Parasites. Academic Press, New York.
KASSAI, T. and I. D. AITKEN. 1967. Indication of immunological tolerance in rats to *Nippostnongylus brasiliensis* infection. *Parasitology* 57: 403-418.
KENNEDY, C. R. 1970. The population biology of helminths of British freshwater fish. *In:* A. E. R. Taylor and R. Muller (eds.), Aspects of Fish Parasitology (Vol. 8). Symposium Brit. Soc. Parasitol., Blackwell Sci. Pub., Oxford and Edinburgh.
KENNEDY, C. R. 1974. The importance of parasite mortality in regulating the population size of the

acanthocephalan *Pomphorhynchus laevis* in goldfish. *Parasitology 68:* 93-101.

KENNEDY, C. R. 1975a. The natural history of Slapton Lay Natural Reserve. *Field Studies 4:* 177-189.

KENNEDY, C. R. 1975b. Ecological Animal Parasitology. Blackwell Sci. Pub., Oxford and Edinburgh.

LEES, E. 1962. The incidence of helminth parasites in a particular frog population. *Parasitology 52:* 95-102.

LESTER, R. J. 1971. The influence of *Schistocephalus* plerocercoids on the respiration of *Gasterosteus* and a possible resulting effect on the behavior of the fish. *Can. J. Zool. 49:* 361-366.

LEWIS, J. W. 1968. Studies on the helminth parasites of the long-tailed field mouse, *Apodemus sylvaticus sylvaticus* from Wales. *J. Zool. Lond. 154:* 287-312.

LI, S. Y. and H. F. HSU. 1951. On the frequency distribution of helminths in their naturally infected hosts. *J. Parasitol. 37:* 32-41.

LIE, K. J., D. HEYNEMAN, and N. KOSTANIAN. 1975. Failure of *Echinostoma lindoense* to reinfect snails already harboring that species. *Internat. J. Parasitol. 5:* 483-486.

MACARTHUR, R. H. and E. O. WILSON. 1967. The Theory of Island Biogeography. Princeton Univ. Press, Princeton.

MCNAUGHTON, S. J. 1975. r- and K-selection in *Typha*. *Amer. Nat. 109(967):* 251-261.

MICHEL, J. F. 1969. The epidemiology and control of some nematode infections of grazing animals. *Adv. Parasitol. 7:* 211-282.

NELSON, G. S. 1959. *Schistosoma mansoni* infection in the West Nile District of Uganda. V. Host parasite relationships. *E. African Med. J. 36:* 29-35.

NEYMAN, J. 1939. On a new class of 'contagious' distributions, applicable in entomology and bacteriology. *Ann. Math. Stat. 10:* 35-57.

NOVAK, M. 1975. Cortisone and the growth of populations of *Mesocestoides tetrathyridia* in mice. *Internat. J. Parasitol. 5:* 517-520.

ODENING, K. 1976. Conception and terminology of hosts in parasitology. *Adv. Parasitol. 14:* 1-93.

ODUM, E. P. 1971. Fundamentals of Ecology. Saunders. Philadelphia.

OLLERENSHAW, C. B. 1959. The ecology of the liver fluke (*Fasciola hepatica*) *Vet. Rec. 71:* 957-963.

OLLERENSHAW, E. P. and L. P. SMITH. 1969. Factors and forecasts of helminthic disease. *Adv. Parasitol. 7:* 283-323.

PARKER, E. D., M. F. HIRSCHFIELD, and J. W. GIBBONS. 1973. Ecological comparisons of thermally affected aquatic environments. *Water Poll. Cont. Fed. 45:* 726-733.

PEARL, R. 1930. The Biology of Population Growth. A. A. Knopf, Inc., New York.

PENNYCUICK, L. 1971. Frequency distribution of parasites in a population of three-spined sticklebacks, *Gasterosteus aculeatus* L., with particular reference to the negative binomial distribution. *Parasitology 63:* 389-406.

PERSIGAN, T. P., M. FAROOQ, N. G. HAIRSTON, J. J. JANREGUI, E. G. GANCIA, A. T. SANTOS, B. C. SANTOS, and A. A. BESA. 1958. Studies on *Schistosoma japonicum* infection in the Phillipines. I. General considerations and epidemiology. *Bull. World Health Organ. 18:* 345-355.

PIANKA, E. R. 1970. On r- and K-selection. *Amer. Nat. 104:* 592-597.

PIANKA, E. R. 1972. r- and K-selection or b and d selection? *Amer. Nat. 106:* 581-588.

PIANKA, E. R. 1974. Evolutionary Ecology. Harper & Row, Inc., New York.

POLYA, G. 1931. Sur quelques points de la théorie des probabilitiés. *Ann. Inst. Henri Poincaré. 1:* 117-161.

ROBERTS, L. S. 1966. Developmental physiology of cestodes. I. Host dietary carbohydrate and the 'crowding effect' in *Hymenolepis diminuta. Exp. Parasitol. 18:* 305-310.

ROGERS, W. P. and R. I. SUMMERVILLE. 1963. The infective stage of nematode parasites and its significance in parasitism. *Adv. Parasitol. 1:* 109-177.

RYSAVY, B. 1966. The occurrence of cestodes in the individual orders of birds and the influence of food on the composition of the fauna of bird cestodes. *Folia Parasit. (Praha) 13:* 158-169.

SCHAD, G. A. 1963. Niche diversification in a parasitic species flock. *Nature, Lond. 198:* 404-407.
SCHAD, G. A. 1966. Immunity, competition and natural regulation of helminth populations. *Amer. Nat. 100:* 359-364.
SCHMID, W. D. and E. J. ROBINSON. 1972. The pattern of a host-parasite distribution. *J. Parasitol. 58:* 907-910.
SEIDENBERG, A. J., P. C. KELLEY, E. R. LUBIN, and J. D. BUFFINGTON. 1974. Helminths of the cotton rat in southern Virginia with comments on the sex ratios of parasitic nematode populations. *Amer. Mid. Nat. 92:* 320-326.
SMYTH, J. D. 1962. Introduction to Animal Parasitology. Eng. Univ. Press., London.
SMYTH, J. D. 1969. Parasites as biological models. *Parasitology 59:* 73-91.
SMYTH, J. D. and M. M. SMYTH. 1968. Some aspects of host-specificity in the *Echinococcus granulosus*. *Helminthology 9:* 519-529.
SOULSBY, E. J. L. 1963. Immunological unresponsiveness to helminth infections in animals. *Proc. 17th World Vet. Congr. 1:* 761-767.
SPALL, R. D. and R. C. SUMMERFELT. 1970. Life cycle of the white grub, *Posthodiplostomum minimum* (Maccallum, 1921: Trematoda: Diplostomatidae), and observations on host-parasite relationships of the metacercaria in fish. *In:* S. F. Snieszko (ed.), A Symposium on Diseases of Fishes and Shellfishes. *Amer. Fish. Soc.*, Washington, D. C. (Spec. Publ. No. 5).
THOMAS, M. 1949. A generalization of Poisson's binomial limit for use in ecology. *Biomet. 36:* 18-25.
VON BRAND, T. 1966. Biochemistry of Parasites. Academic Press. New York.
VON BRAND, T. and T. I. MERCADO. 1961. Histochemical glycogen studies on *Fasciola hepatica*. *J. Parasitol. 47:* 459-461.
WEINLAND, E. and T. VON BRAND. 1926. Beobachtungen and *Fasciola hepatica*. *Vergl. Physiol. 4:* 212-285.
WILLIAMS, H. H. 1960. The intestine in members of the genus *Raja* and host-specificity in the Tetraphyl-

lidea. *Nature Lond.* *188:* 514-516.
WISNIEWSKI, W. L. 1958. Characterization of the parasite fauna of an eutrophic lake (Parasitofauna of the biocoenosis of Druzno Lake - Part I). *Acta Parasit. Pol.* *6:* 1-64.
YUSUFI, A. N. K. and A. H. SIDDIQI. 1976. Comparative studies on the lipid composition of some digenetic trematodes. *Internat. J. Parasitol.* *6:* 5-8.
YAMAGUTI, S. 1958. Systema Helminthum. Interscience Pub., Vol. II. New York.

The Regulation of Fish Parasite Populations

C. R. KENNEDY

*Department of Biological Sciences
University of Exeter
Exeter, United Kingdom*

INTRODUCTION

The theme of this volume is the regulation of parasite populations, and this contribution is concerned specifically with the regulation of fish parasite populations, regardless of whether the fish serves as the definitive or the intermediate host for the parasite. In particular, it will try to answer the question of how far fish parasite populations can be, and are, regulated at all, and the extent to which regulation occurs at the stage of their life cycle when they are inhabiting a fish and involves the fish-parasite interaction. It is essential to remember, however, that regulation of a parasite suprapopulation (see Esch, Gibbons, and Bourque (1975) for a discussion of this, and the term, infrapopulation) can occur at any stage or within any host, whether an invertebrate intermediate host or another fish definitive host species, in the parasite's life cycle. This aspect of parasite population biology and the relationship between population regulation and community organization will not be considered in any detail here, as it is discussed fully in this volume by Holmes.

REGULATION AND STABILITY

CONCEPT OF STABILITY

Before embarking upon a discussion of regulatory factors and the extent to which fish parasite populations are stable or unstable it is essential to clarify the terminology employed. The growth of a fish parasite population does not differ in any fundamental way from the growth of a free-living animal population. If unchecked, fish parasite population growth would be exponential. This exponential growth will continue until constraints start to operate upon it. The constraints may operate equally effectively over the whole range of population densities, in which case they are density-independent, or they may operate with increasing severity as the population size increases, in which case they are density-dependent or, in systems terminology, negative feedback controls. Both types of constraint may operate upon a population, and they may affect the basic population processes of birth rate (natality), death rate (mortality) and immigration (recruitment) rate.

Although parasite populations may persist in any habitat for longer or shorter periods of time, this does not necessarily mean that they are stable. Stability is used here in the sense of May (1973) and Anderson (1974a): it refers to the ability of any population or system to come to an equilibrium level, and to return to that equilibrium level or to a new equilibrium level, and to return to that equilibrium level or to a new equilibrium level if perturbed or displaced. Persistence of a population is thus not *per se*, evidence of stability. The equilibirum may be static, but is far more often dynamic and involves oscillations about a mean level. A stable population or equilibrium can only be achieved by the operation of density-dependent factors (negative feedback controls) or regulatory processes, and so regulation is used here to refer to the ability of a population to attain stability by the employment of such factors or controls. An unstable population is one that is unregulated and so is in constant danger of extinction; it is constrained only by density-independent factors. This concept of stability does not imply that in a stable parasite population all individual host-parasite interactions are stable.

It has often been assumed that because parasite life cycles are very complex, involving in some cases as many as six infrapopulations of parasites, e.g. eggs, miracidia, sporocysts, cercariae, metacercariae and adults, and at least three host populations, this complexity alone is sufficient to impart stability to the parasite population. This is not the case. May (1973) has shown convincingly that in model ecosystems complexity alone is not sufficient to ensure stability, and that increasing complexity may even decrease stability. This has been confirmed in respect of parasite populations in the models of Anderson (1974a; in press), who has gone on to show that the complexity of parasite life cycles is important in that it increases the number of stages or places at which density-dependent controls can act. The more complex the life cycle, the more the opportunities for such action. Nevertheless, even though several controls may operate at different stages in the life cycle, Anderson (in press) has also demonstrated that only a single negative feedback control mechanism operating in a single host species can be sufficient to regulate the whole parasite suprapopulation. Complexity of life cycle is important in another respect also. Both Anderson (in press) and May, et al. (1974) are agreed that the more complex the life cycle, the more opportunities there are for time lags in the system, and that, in general, time lags destabilize the system.

REGULATORY MECHANISMS

In a discussion of parasite population stability, Bradley (1974) recognized three basic types of regulation of parasite population numbers. His first type was by transmission of the parasite, as he recognized that small changes in transmission rate could lead to large changes in population size. Unless, however, transmission rate is density-dependent it cannot actually regulate a population size in the sense that regulation is used here. Such evidence as there is to date suggests that transmission rates of parasites are not density-dependent, and that larval survival is age-dependent and decreases exponentially with time (Anderson and Whitfield, 1975; Kennedy, in press). Transmission rates may, and often do, determine the size of the population and the equilibrium level (Crofton, 1971), but for stability and regulation the operation of density-dependent factors in addition is essential. Bradley's Type I mechanism cannot, therefore, actually regulate parasite populations.

Regulatory mechanisms may operate upon birth, death, or immigration rates, and of those found to operate upon parasite populations, many also operate upon free-living animals. Intraspecific competition or crowding (Bradley Type III), for example, is known to occur amongst free-living and parasitic animals, and may affect growth rate, and so generation time, parasite size, and so egg production, and parasite survival, and so mortality. Two other regulatory mechanisms, however, are particularly applicable to, or virtually confined to, parasites. The first of these involves regulation by death of the host (Bradley Type II). Crofton (1971) has demonstrated in model systems that regulation of both host and parasite populations can be achieved if parasites are overdispersed and are able to kill their hosts, and that both host and parasite populations can then reach a stable equilibrium. Both the lethal level and the degree of overdispersion depend upon the number of parasites, so the mechanism is density-dependent. This type of regulation will be discussed in more detail later.

The second regulatory mechanism, which is confined to parasitic animals, involves immune or other responses by the host (Bradley Type III). The extent or degree of these responses is almost always dependent upon the degree of stimulation, i.e. the parasite numbers, and so the mechanism is density-dependent. The manifestation of host immune responses includes increased parasite mortality, decreased fecundity, increased generation time and reduced establishment in subsequent infections.

Any one or all of these regulatory mechanisms can act at any or all the parasitic stages in the life cycle of a parasite, and so do not necessarily act upon fish parasites when they are present in their fish host. The behavior of the parasite population at this stage in its life cycle could easily be determined by the behavior of an infrapopulation at an earlier stage and in an intermediate host.

SOURCES OF EVIDENCE FOR REGULATION OF FISH PARASITE POPULATIONS

It must be emphasized at the outset that evidence pertaining to the regulation of fish parasite populations is sparse, and often comes from studies carried out for other purposes. In comparison with the situation amongst free-living animals and parasitoid insects (Varley, Gradwell, and Massey, 1974), there are few studies of fish parasite populations that have had population regulatory mechanisms as their specific aim.

The evidence for regulation amongst fish parasite populations comes from three main sources. The first of these is general models of host-parasite systems, such as those of Anderson (in press) and Crofton (1971), which by their nature apply also to fish-parasite systems. Such models of course tend to show how regulation can happen, rather than how it does in any particular situation. The second source is long-term studies on changes in fish parasite population levels extending over several years. There are very few such studies, and fewer still that have also looked at the contemporary changes in parasite infrapopulations in other hosts. The third source is short-term studies, generally of Ph.D. length, on seasonal changes in parasite infrapopulations in fish. These seldom consider the parasite in its free-living stages or in its other hosts. Correlations between the changes in parasite infrapopulations in fish and climatic factors and/or fish behavior and diet are often noted, but there have been only a few experimental investigations into the causal relationships. In some instances mathematical models have also been constructed in an attempt to understand and verify the particular relationships observed in the field. These short-term studies are, unfortunately, by far the most frequent, and so supply the greater part of the information available about regulation of fish parasite populations.

OPERATION OF DENSITY-INDEPENDENT FACTORS

Very many species of parasite when in their fish host exhibit seasonal cycles in incidence and intensity of infection and in maturation. Such cycles have been reported for monogeneans (Paling, 1965), digeneans (Wootten, 1973), cestodes (Kennedy and Hine, 1969), acanthocephalans (Muzzall and Rabalais, 1975), copepods (Tedla and Fernando, 1970), and leeches (Halvorsen, 1971) in freshwater fish, and in marine fish (Llewellyn, 1962; Mackenzie and Gibson, 1970; Boxshall, 1974a; 1974b), to give only a few examples. They almost certainly occur also in protozoans, although these have not been studied in such detail, and in regions other than the sub-arctic and temperate latitudes. The occurrence and causes of seasonal cycles have been reviewed by Kennedy (1970; 1975), and it is concluded that most can be correlated with changes in fish behavior and diet, and directly or indirectly with physical changes in the water, and especially with temperature.

Although such correlations have been noted on many occasions, in only a few cases have the dynamics of the parasite populations been studied in any detail and attempts made to determine whether the relationship with temperature is a causal one, and if so whether it affects parasite natality, mortality or immigration and whether directly or indirectly. From such studies as have been carried out, it is clear that the effects of density-independent factors are very variable in time and space, and affect different host-parasite systems very differently. This is particularly well exemplified by the proteocephalid cestodes. Nearly all the species investigated exhibit well-defined seasonal cycles in incidence and maturation, with, in the Northern Hemisphere, maturation occurring in spring, and the adult population dying out soon after or reaching a very low level in summer (Fischer, 1967; Fischer and Freeman, 1969; Hopkins, 1959; Kennedy and Hine, 1969; Wootten, 1974). The population dynamics of each species may, however, differ considerably. In Glasgow, Scotland, *Proteocephalus filicollis* shows a clear seasonal cycle (Hopkins, 1959), with immigration into the fish taking place from June until November, when infective copepods are no longer available. The parasites grow mainly in autumn and spring, mature in June and July and die soon after. Immigration and natality are thus seasonal, but mortality, other than that due to the death of spent adults (1%), is constant throughout the year. In Yorkshire, however, all developmental stages of the parasite were present in fish at all times of the year (Chappell, 1969a). Infected copepods were available and eaten all through the winter, and immigration appeared to be greatest between November and January. Gravid adults were also found all year, but were most common in August and September. The main cause of turnover was the death of spent adults, which thus showed a seasonal cycle, and was again density-independent.

Working in Ontario, Fischer and Freeman (1969) showed that the incidence of *Proteocephalus ambloplitis* in the intestine of smallmouth bass was seasonal, being highest in spring, declining in summer, and disappearing completely by late autumn. Plerocercoids were found in the viscera all year, and they were able to show experimentally that a rise in water temperature from $4°C$ to $7°C$ or higher stimulated the parenteral forms to leave the viscera and penetrate into the intestine. This rise in temperature took place in the lake in May and June, and thus accounted for the seasonal appearance of the parasites in the fish intestine. In Gull Lake, Michigan (U.S.A.), *P. ambloplitis* also appeared in bass intestines in May, but at that time the water temperature was $14°C$, the rise to $7°C$ having taken place five weeks previously (Esch, Johnson, and Coggins, 1975). Eure and Esch (1974) reported that in Aiken, South Carolina (U.S.A.), where the lowest annual water temperature was $8°C$, adult parasites were present in the intestine of bass in January. Thus, although a rise in temperature may cause seasonal immigration, other additional factors are clearly involved.

Density-independent factors also appear to affect populations of *Triaenophorus nodulosus* in different ways in different localities. The parasite infrapopulation in the fish intermediate host, *Perca fluviatilis*, is in a state of dynamic equilibrium between loss and gain of parasites. Plerocercoids are acquired throughout the year, but especially in spring. Mature plerocercoids infective to the definitive host, *Esox lucius*, are available all year, but they live for only 2 - 3 years in perch and die mainly in summer. Older perch are thus not more heavily infected than younger ones, as death of the parasites then approximately equals immigration into the fish. This cycle is the same in Britain and Norway (Chubb, 1964; Lien, 1970). The adults become gravid in spring in both countries and in Canada (Miller, 1943; Chubb, 1963; Borgstrom, 1970), but in Britain plerocercoids invade pike all year, and there is no increase in intensity of infection to a maximum at any time of year. The parasite infrapopulation is thus in a state of dynamic equilibrium between loss and gain of parasites (Chubb, 1963), but the mechanism relating immigration to mortality rates is not known. In Norway, by contrast, intensity of infection in pike shows a seasonal change, with low levels in winter, a maximum in May and a subsequent sharp decline, with parasite levels remaining low, over summer. Pike, however, eat more perch in summer and presumably acquire more plerocercoids. Loss of parasites must thus be high in summer and low in spring, when the infrapopulation level increases. Mortality rates at least must be changing seasonally, and the dynamic equilibrium situation found in Britain certainly does not appear to occur in Norway (Borgstrom, 1970).

In all these examples the differences in the population changes in the fish parasites from site to site are completely consistent with the operation of density-independent factors, but in none have the investigations been sufficiently detailed to show that density-dependent factors are not also operating. In only three cases has this been convincingly demonstrated. The first of these concerns the parasitic copepod *Lepeophtheirus pectoralis* on plaice *Pleuronectes platessa*. Boxshall (1974a; 1974b) has shown that the population dynamics of this parasite are causally related to temperature cycles. The parasite has two generations a year an overwintering generation which breeds in May, and a shorter lived summer generation which breeds in August and gives rise to the winter generation. The parasites die after breeding, from natural senescence. The parasites are overdispersed

and occur on the fish in clumps, and mortality within
a clump is independent of the clump size, i.e. is
density-independent. The parasite does not cause the
death of the fish, and re-infection of fish is common.
The aggregation of the free-swimming larvae in the sea
is responsible for the overdispersion of the parasites
on the fish and the clump size depends on the immigration process. Changes in the infrapopulation size of
the copepods on plaice relate mainly to changes in the
total number of clumps, which in turn relate to the infection process. Thus natality, mortality and immigration are all density-independent, and population levels
are determined by factors such as the water temperature, the fish shoaling and migration behavior, and the
effects of physico-chemical conditions upon the behavior of the infective larvae.

The second case is that of the monogenean gill
parasite *Diplozoon paradoxum*. This species also shows
a seasonal cycle of incidence and intensity of infection, growing in spring, breeding in June and with the
greatest mortality in summer when the water temperature
is warmest. Most immigration also takes place in summer, but the parasite lives for two years or more and
so, as it is present on fish at all times of year, the
seasonal cycles are not so obvious (Halvorsen, 1969;
Anderson, 1974b). Nevertheless both authors have
clearly shown that the seasonal variation in population
size is due to seasonal trends in immigration and death
rates, which are in turn temperature-dependent. Anderson (1974b) has further demonstrated that there is a
close relationship between host age, and thus size, and
the number of *D. paradoxum* on any fish. This is associated with the space available for the parasite in the
gill chamber, the volume of water passing over the gill
filaments and carrying the infective stages, and the
accumulation of parasites with time. In small fish the
parasites tended to be underdispersed on the gills and
to have a more regular distribution than in older ones.
Anderson suggested that this was due to competition for
space amongst the parasites in a habitat of restricted
dimensions. In older fish, where the density of infective stages was not sufficient to saturate the gill
filaments or reach the carrying capacity of the gills,
the parasites were overdispersed, thus reflecting the

heterogeneity of infection experiences amongst the fish. Thus, the major factors affecting the parasite population size and dispersion were the immigration process and the size of the habitat. Immigration was density-independent, but competition could of course function in a density-dependent manner, although Anderson believed that the carrying capacity of the gills was often not reached.

The third case is that of the cestode *Caryophyllaeus laticeps*. Kennedy (1968; 1969) has shown that there is a very pronounced seasonal cycle in the infrapopulation size in fish. Infection commenced in December, immigration continued until May and then ceased until the next winter. Loss of established parasites from fish commenced in February and continued until May and June, and between July and December the parasite was completely absent from fish. The parasites produced eggs only in April and May. This pattern of cyclical changes was observed in this locality for a period of four years (Kennedy, 1972b). Although immigration was clearly seasonal, it was not due to seasonal availability of infective larvae, since these were present in the intermediate host well before and well after the period of recruitment into the fish (Kennedy, 1969). It was suggested that seasonal immigration was related to the feeding habits of the fish. The whole cycle showed a strong negative correlation with water temperature changes, and Kennedy (1971) was able to demonstrate a causal relationship between water temperature and parasite mortality. As water temperature rose from winter to summer levels, parasites were rejected more rapidly from the fish and new infections could not establish. The rejection did not involve a host immune response (Kennedy and Walker, 1969). The population level was thus controlled by seasonal changes in immigration and mortality, both density-independent factors, and the latter at least causally related to temperature changes.

In another locality Anderson (1974a; 1974c) found a very similar pattern of seasonal population changes in fish, but extended the study to include a wider range of fish age groups. He found seasonal periodicity in all age groups, but with different phases and amplitudes of cycles, and was able to relate this to the different feeding habits of the age classes, which changed seasonally. He thus confirmed that seasonal immigration was due to changes in fish feeding habits, and that seasonal mortality was related to rising temperatures and so was density-independent. Extinction of the suprapopulation in summer when adults disappeared from fish was prevented by the ability of infrapopulations to survive within intermediate hosts during this period. Anderson also suggested that the seasonal maturation might assist in regulating the population size, since the maximum number of eggs were produced when the infrapopulation size in fish was lowest, thus preventing too large a proportion of intermediate hosts becoming infected. Assuming that the infrapopulation size in fish was determined by the immigration - death processes and that seasonality resulted from the combined influence of these factors, Anderson then constructed a model of the system. The model proved to fit the empirical data very well. Using a deterministic model, he showed that the adult population exhibited damped oscillations tending towards a stable equilibrium, or reached a stable oscillating equilibrium. In a stochastic model the number of adults oscillated cyclically but never reached a steady state due to chance fluctuations. Overdispersion could aid regulation (as predicted by Crofton, 1971) but was not essential. Stability was enhanced if the death rate of the hosts was proportional to the number of parasites, but in natural situations, *C. laticeps* has not been shown to be pathogenic. Population levels were particularly sensitive to changes in immigration and birth rates of parasites, and to changes in host population sizes. Natural variations in water temperature from year to year or from locality to locality did not markedly affect the stability of the cyclic behavior, however, which has been confirmed since by the field studies of Milbrink (1975).

    Density-independent factors therefore have been shown to have very pronounced effects upon parasite infrapopulation changes in fish. Despite the large number of observations and correlations between population changes and factors such as water temperature, in only a few cases has it been demonstrated precisely how these factors operate. It is also clear from these cases that a similar pattern of cyclical oscillations in parasite numbers can persist over several years, controlled entirely by density-independent factors. These populations must nevertheless be regarded as unstable, and liable to extinction as a result of unfavorable chance fluctuations in climatic parameters.

OPERATION OF DENSITY-DEPENDENT FACTORS

REGULATION BY HOST DEATH

The regulation of parasite populations by death of heavily infected hosts has been demonstrated by Crofton (1971) in a series of general models of host-parasite systems. He represented the reproductive rate of the parasite and its potential to infect a host by an Achievement Factor, Af. He assumed that all parasites were overdispersed throughout their host population and that the dispersion could be described by the negative binomial model. The parameter k of the negative binomial model could be used as a measure of the overdispersion, such that at high values of k, overdispersion was low and the distribution tended towards randomness. He also assumed that all parasites were capable of killing their host at a certain level of infection, the Lethal Level, designated by L. Using deterministic models, and considering these factors only but recognizing that a variety of other density-dependent and density-independent factors acted upon all systems, he proceeded to examine the changes in host and parasite numbers over a number of generations at different levels of Af, k and L. His models showed that the numbers of hosts and parasites could attain a stable equilibrium, which could be constant or involve regular cyclic oscillations. Differences in the initial numbers of parasites and hosts only affected the time taken to attain the equilibrium level and not the equilibrium level itself. As Af increased, population levels fell and the time taken to reach equilibrium increased. As L increased (pathogenicity decreased), equilibrium levels were higher but the time taken to attain stability increased. As long as k remained between one and three, the system was stable and regulated. Sterilization of the host as a result of the parasite was equivalent to death in population terms, and so could be considered with changes in L. Crofton also considered the effects of immune responses, but these will be discussed later. In essence, what Crofton demonstrated was that under the conditions he laid down, host and parasite populations could be regulated and stable, and that this regulation involved the ability of the parasite to kill its host. It depended also upon the degree of overdispersion of the parasite, since for regulation it was essential that the death of a heavily infected host should remove a greater proportion of the parasite population from the system than of the host population. It was one way in which parasite populations could be regulated, and could in theory apply to all parasite populations provided they were able to kill their hosts at some level of infection.

What is not always generally appreciated, however, is that Crofton's models are just as informative about the conditions that produce instability and unregulated systems. For example, he showed that as Af decreased and approached zero, the host population size increased proportionately and its growth tended to become exponential; similarly, as Af increased, not only did the population levels fall, but also the whole system tended to become unstable. As pathogenicity decreased and the level of L increased, the system also tended to become unstable. Finally, when k was greater than three, the oscillations of host and parasite numbers increased in amplitude and instability, and often became extinct, whereas when k was less than one no regulation occurred, or at least if the populations were regulated, it was by factors other than those considered in the model. In one simulation, when k=5, Af=6 and L=10, the whole system went into prolonged and violently unstable oscillations, resulting eventually in its breakdown. In addition to the models actually showing instability, some of the assumptions made by Crofton in constructing them are also revealing of conditions for instability. The models, for example, are deterministic, yet many of the variables should be treated stochastically (and especially L). Crofton himself recognized this, and pointed out that L would depend upon the condition of the fish and factors other than just the parasite numbers. In all the models k was treated as a constant, whereas in fact it can and does vary seasonally (Boxshall, 1974a; 1974b; Pennycuick, 1971c). Only those variables specifically stated are considered in the models, and no account is taken of the wide variety of climatic and other density-dependent factors than can affect parasite population levels. Finally, Crofton assumed that all parasites in hosts containing more than the lethal level were eliminated from the system. This may be true when the parasites are in their definitive host, but the removal of a heavily infected intermediate host from the system may not remove the contained parasites, since these may still go on to infect the definitive host and eventually produce offspring.

Crofton's models should thus be regarded as showing what can happen under the conditions laid down and when a parasite can cause host mortality. They indicate one way in which host and parasite populations can interact and the parasite population be regulated. They make no assumptions about the frequency with which this method of regulation is actually encountered in natural situations. Equally, they indicate clearly conditions under which instability and unregulated populations might be expected to occur. Since, however, many fish parasites are known to have deleterious effects upon their hosts and to cause host mortality, it is important to consider whether any fish parasite infrapopulations are regulated in the way Crofton predicted.

In a detailed study of the parasites of the threespined stickleback, *Gasterosteus aculeatus*, Pennycuick (1971a; 1971b; 1971c; 1971d) was able to show that *Schistocephalus solidus*, the plerocercoids of which live in the fish body cavity, was responsible for host mortality. The major cause of the mortality was predation on heavily infected fish, and this was highest in winter when the climate was more unfavorable. There was some evidence that the major cause of mortality of *Diplostomum gasterostei* metacercariae was also due to death of heavily infected fish, whereas the major mortality in *Echinorhynchus clavula* was the death of individual parasites. All three species were overdispersed throughout the stickleback population, but only *E. clavula* and *D. gasterostei* could be described by the negative binomial model. For *Schistocephalus solidus*, a log normal provided a better fit. In all cases, this resulted in a large number of parasites occurring in a small number of hosts, such that the death of a fish from predation reduced the parasite population more than that of the host and kept the infection to a moderate level. The overdispersion was highest for *D. gasterostei* and lowest for *Schistocephalus*, although this latter species had the greatest pathological effect on the fish. *E. clavula* was not lethal. Pennycuick believed that regulation of *Schistocephalus* at least was due to its causing host death. It also reduced the weight, condition factor and growth of the fish, delayed sexual maturation and rendered many of them sterile. She provided evidence for cyclic fluctuations in the numbers of sticklebacks and *Schistocephalus* and explained this in terms of Crofton's (1971) predictions. As infection levels rose, more sticklebacks died and the population became smaller. As the fish population declined, so did the success of the parasite in locating new hosts, and so its population level also fell. At a certain level, it no longer exerted any significant effect on the fish population, which was then able to build up again and so the cycle was repeated. The intensity of regulation depended upon the degree of overdispersion, and there was clearly an optimum level of infection, since if parasite levels were too low, the heavily infected fish would not be found by predators, and if they were too high, too many fish would die before they were eaten. The period of oscillation would be influenced by fish diet and water temperatures, as well as by the levels of *D. gasterostei*, which was also lethal. Thus, the *Schistocephalus* and *D. gasterostei* populations and that of the sticklebacks appeared to be regulated by parasite-induced, host mortality in the manner predicted by Crofton.

Schistocephalus has been studied in many other localities (Lester, 1971; Chappell, 1969a; Arme and Owen, 1967; Walkey and Meakins, 1970; Meakins, 1974; Vik, 1954), and in all cases has been shown to exert similar effects upon sticklebacks. It is always overdispersed, and it always affects fish respiration and growth, and frequently, though not inevitably, fish maturation. In all localities, though, it affects the fish behavior and renders them more susceptible to predation. Arme and Owen (1967) also found some indication of changes in infection levels over the period of study. It seems likely therefore that the *Schistocephalus* and stickleback populations could behave and be regulated in a similar way to that demonstrated by Pennycuick.

Many other species of fish parasite have been reported to be lethal, and the effects of fish parasites on their hosts have been reviewed on a number of occasions (Bauer, 1961; 1962; Bauer, Musselius, and Srelkov, 1973; Petrushevski and Shulman, 1961; Rees, 1967; Williams, 1967; Arme and Walkey, 1970). Most of these authors are agreed, however, that the majority of fish parasites are not lethal under natural conditions, but only under the somewhat unusual conditions of fish farms. The parasites that are capable of killing their hosts under natural situations are nearly always larval stages, especially larval cestodes, which use the fish as an intermediate host.

In addition to *Schistocephalus*, there are four species of cestodes that appear to be capable of being regulated by host mortality. Plerocercoids of *Ligula intestinalis* have pronounced effects upon the behavior of infected fish (Orr, 1966; Dence, 1958; Arme and Owen, 1968) which renders them particularly susceptible to avian predation (Van Dobben, 1952). It also renders infected fish sterile, by affecting the production of gonadotropic hormones by the pituitary gland (Kerr, 1948; Arme, 1968). Plerocercoids of *Diphyllobothrium* spp. have also been reported to be responsible for fish mortality on a number of occasions (Hickey and Harris, 1947; Fraser, 1960; Hoffman and Dunbar, 1961; Becker and Brunson, 1967; Halvorsen, 1970). These reports must in general be treated with a certain amount of caution, however, since very often more than one species of parasite is involved, and the locations are frequently reservoirs or managed fisheries where stocking of new fish is of regular occurrence. As Halvorsen (1970) has pointed out, any mortality due to *Diphyllobothrium* in natural lakes is rare, and mass mortality of the type reported in many of these publications is particularly rare. There are also a few reports that

Triaenophorus nodulosus is capable of causing fish mortality (Lopukhina, et al., 1973; Kuperman, 1973) and also *Cyathocephalus truncatus* (Vik, 1954; 1958), but this only appears to occur in some host species and under rather unusual conditions.

Amongst other groups of parasites, larval digeneans of the genus *Diplostomum* are also reported to cause fish mortality (Bauer, 1961; Petrushevski and Shulman, 1961; Davies, Burkhard, and Hibler, 1973; Betterton, 1974; Sweeting, 1974). It is clear from these reports, however, that most cases of mortality have occurred in fish farms, and then only in a few species of fish. Under natural conditions most fish are unaffected by the parasite, although both Pennycuick (1971a; 1971b; 1971c; 1971d) and Davies, Burkhard, and Hibler (1973) have shown that mortality due to this can sometimes occur. Most larval nematodes have little or no effect upon their fish hosts, or if they do exert an effect, it is only upon certain species of fish and in a few cases (Locke, et al., 1974; Kennedy and Lie, 1975).

In general therefore, it appears to be only larval parasites of a few groups, using fish as their intermediate hosts, that cause host mortality. The mortality is seldom direct, but is due more often to infected fish having altered behavior patterns that render them more susceptible to predation. Such a behavioral change is obviously of advantage to the parasite (Holmes and Bethel, 1972). Those species that do cause mortality would appear to be capable of having their populations regulated in the manner predicted by Crofton (1971) and demonstrated by Pennycuick (1971a; 1971b; 1971c; 1971d). However, in no case has long-term regulation or stability been demonstrated. Pennycuick's studies lasted only for two years, and a few years later the pond dried up during a hot summer (Avery, pers. comm.), causing extinction of the host and parasite populations (and demonstrating very effectively the importance of density-independent factors in the system). *Ligula intestinalis* has been shown to cause a long-term decline in a fish population (Wilson, 1971), but no cyclic oscillations of host and parasite were observed, and indeed the lake was stocked with fish throughout the period of decline. Nearly all the reports of *Diphyllobothrium* induced mortality refer to epidemics. Again the long term changes in host and parasite numbers were not recorded, nor have there been subsequent reports of new epidemics from the same localities. Where the effects of *Diphyllobothrium* have been followed over several years the changes have not been oscillatory in nature. Powell and Chubb (1966) showed that in one lake parasite numbers declined in one species of fish over seven years but not in another species, in which they remained constant. Thus, although over short periods many parasite populations appear to be regulated, or to be capable of being regulated, by host death, there is no population in which long-term regulation by this method has been demonstrated, nor have regular oscillations of host and parasite numbers, as predicted, actually been demonstrated. This may be due to the short term over which many of the studies have been carried out, but it may also be due to (1) the limited conditions under which Crofton (1971) showed regulation to operate are seldom encountered in nature, (2) density-independent factors play a more important part than he assumed and (3) in natural situations, the conditions that he predicted for instability are far more frequent.

REGULATION BY HOST RESPONSES

Immune Responses

Although many mammals and birds are capable of mounting effective immune responses against their parasites (Kennedy, 1975), very few fish appear to be capable of doing so. Immune responses in this context refers to responses in which host antibodies, whether circulating or cell-bound, have been conclusively demonstrated to be present and operative against the parasite. The apparent inability of most fish to respond immunologically to their parasites may well be a reflection of the evolutionary state of development of fish immune responses in general. Although most species investigated have been shown to be capable of producing circulating antibodies in response to injected antigenic materials, only IgM and IgG have been identified in fish; IgA, IgD and IgE production has never been demonstrated, and fish appear to be incapable of producing skin-sensitizing antibody or responding with systemic anaphylactic shock (see Harris(1973a; 1973b) for a review of this topic). Furthermore, the production of antibodies by fish is temperature-dependent, although they are produced over the normal temperature range (Harris, 1973b). Fish are thus apparently incapable of producing the wide range of immunological responses characteristic of mammals. Although fish immunoglobulins have been detected in both serum and mucus (O'Rourke, 1961; Fletcher and Grant, 1969; Harris, 1972), their origin and the method by which they get into the mucus is not known, and there is no evidence that they are as effective as the mucoantibodies of mammals.

Serum factors, possibly but not yet proven to be antibody in nature, have been shown to be effective in relation to the specificity of fish parasites (Orr, Hopkins, and Charles, 1969; McVicar and Fletcher, 1970; Arme and Walkey, 1970). In only two cases, however, has the production of antibodies by fish in response to the presence of their normal and specific parasites been demonstrated. Molnar and Berczi (1965) showed the presence of antibodies in the serum of *Abramis brama* infected with plerocercoids of *Ligula intestinalis*, and Harris (1972) showed that *Leuciscus cephalus* produced circulating antibodies in response to excretory-secretory products of mature *Pomphorhynchus laevis*. In neither case were the antibodies demonstrated to have any effect on the parasites eliciting them or to confer immunity to re-infection. In many other cases, even where fish have been shown to mount an effective response against parasites which involve the fish mucus (see next section), antibodies, although searched for, have not been found (Orr, Hopkins, and Charles, 1969; Kennedy and Walker, 1969; Lester, 1972; Nigrelli, 1935; 1937).

To date, therefore, there is no evidence that classic immune responses involving antibodies are effective against any species of fish parasite. Despite the demonstration by Crofton(1971) that immune responses in conjunction with parasite induced host mortality can regulate the size of parasite populations, there is a general agreement (Bauer, 1962; Rees, 1967;

Williams, 1967; Kennedy, 1975) that immune responses play no part in the regulation of parasite infrapopulations in fish.

Other Responses

In a few cases it has been demonstrated that fish can mount effective responses against their parasites. The responses are all directed against ectoparasites on the skin or gills, they involve the fish mucus, they confer partial or complete protection against re-infection for a longer or shorter period, and the precise mechanism is not known in any case. Until it is shown conclusively that an immune response is involved, it appears preferable to consider them in a separate category.

Although a response to glochidia larvae involving hypertrophy of gill epithelial tissue and mucus production and conferring some degree of protection against re-infection, has been reported (in Bauer, 1962), the first convincing demonstration of an effective response to fish parasites was given by Nigrelli (1935; 1937). Heavy infections of *Epibdella (Benedenia) melleni* on the eyes of several species of marine fish in the New York Aquarium killed the fish. If the fish survived the infection, however, parasite numbers declined, re-infection at the same site was not possible, and any new infections settled at different sites. The mechanism was shown to involve the fish mucus, which became toxic to parasites following infection and rendered the site of previous infections unsuitable for *E. melleni*. Attempts to immunize the fish were not successful, and despite suggestions to the contrary (Jahn and Kuhn, 1932), the involvement of humoral immunity in the response could not be demonstrated.

Local host responses to species of *Dactylogyrus* involving proliferation of gill epidermis and excessive mucus production and conferring some degree of protection against re-infection, have been reported on several occasions (Wunder, 1929; Bauer, 1962; Putz and Hoffman, 1964; Paperna, 1963; 1964). They have been studied in greatest detail by Paperna (1963; 1964), however, and particularly in relation to mixed-infections of *Dactylogyrus* species on the gills of *Cyprinus carpio*. Paperna noted that in fish ponds, carp infected with *D. vastator* were only lightly infected or not infected at all with *D. extensus*. In the laboratory, fish harboring *D. extensus* and exposed to infections with *D. vastator*, lost their *D. extensus*, which was replaced by *D. vastator*. This loss was associated with histological changes in the gills, which eventually rendered them unsuitable for, and resulted in the loss of, *D. vastator* itself. After a period of healing, the gills became suitable for re-infection. Eventually, however, specific resistance to *D. vastator* developed which eliminated the species and left the fish completely refractory to re-infection by this species, although the gills remained suitable for *D. extensus* and other species. Infections of *D. vastator* were thus virtually self-terminating, as a result of a specific host response involving hyperplasia of the gill epithelium and mucus production, which occurred subsequently to a non-specific cell response that caused all parasites to leave.

A very similar general cell response by fish gills to infections with species of *Gyrodatylus* was observed by Khalil (1964), but has been studied in great detail by Lester (1972) and Lester and Adams (1974a; 1974b) in infections of *G. alexandrei* on *Gasterosteus aculeatus*. These latter authors have further gone on to consider the effects of the response on the *Gyrodactylus* population as a whole, rather than just to consider the effects on the parasites within an individual fish. Lester (1972) showed first that fish naturally or experimentally infected with *G. alexandrei* lost their infections within four weeks. Sticklebacks normally shed a layer of mucoid material (slough) every one - two days, and since the parasites are attached to the slough and not to the dermis, they are shed with it. The slough is not a reaction to the parasite, but increases in density in association with infections of *Gyrodactylus*, stimulated probably by the parasite's feeding or attachment. Fish re-infected within a few days of recovery from an infection of *G. alexandrei* lost all their new infections within a week. If kept for four weeks before re-infection, however, the new infections established and repeated the normal cycle of infection. Protection against re-infection was thus of short duration only. In fish recovering from an infection, no increase in the number of mucus cells was found, and it was concluded that a humoral immune response was not involved since no long-term immunity could be demonstrated, the response was very rapid, and injection of antigens conferred no protection. The response thus differs from the specific host response to *Dactylogyrus vastator*.

In an attempt to examine the effects of this response on the whole infrapopulation of *G. alexandrei*, Lester and Adams (1974b) constructed a model of the host-parasite system. They first showed that heavily infected fish (with over 150 parasites) died as a direct result of the infection. They then calculated life tables for the parasite, measured factors affecting its abundance, and demonstrated that the rate of loss of parasites was the same in both light and heavy infections, i.e. the mortality was not density-dependent. Their first model was a simple deterministic one, describing changes on isolated fish. This did not fit the experimental data very well, as it showed parasite numbers declining to low values but persisting, rather than completely disappearing, as had happened in the experiments. They then changed to a stochastic model, which proved a better fit. The model was then enlarged to incorporate several hosts, some recovering and some susceptible to infection. Using a deterministic model, they found that over 10 months, the numbers of parasites oscillated regularly (between 10 and 40 per fish), in a manner similar to that predicted by Crofton's immune model (Crofton, 1971). In the stochastic model, regular oscillations disappeared. Although the parasite numbers on individual fish continued to show fluctuations, they were out of phase. Large fluctuations occurred at irregular intervals, and in 20% of the simulations the population of parasites became extinct. The model departed somewhat from the real situation in that high numbers and resulting fish death were not included. Perhaps as a consequence of this, the long-term predictions based on the model could not be confirmed experimentally. Lester and Adams therefore concluded that in a heterogeneous natural fish population, parasite numbers would fluctuate cyclically on individual fish, but the total infrapopulation numbers would remain fairly constant. The life cycles of both the fish and the parasites were temperature-dependent, and so parasite numbers would build up in winter and spring, but would decline in summer as fish over-responded to their presence. Short-term temperature fluctuations or periods of favorable temperatures could result in an epidemic. Parasite numbers were regulated by the response of the individual fish since in its absence or when it broke down, fish and their parasites would die, and when it was effective fish carried only on average 20% of the lethal level of parasites. The parasite population was still, however, strongly influenced by density-independent factors, especially temperature, and there remained some doubt as to whether the regulatory mechanism could therefore produce long-term stability of the parasite population.

Effective host responses conferring long-term protection against re-infection have been demonstrated in carp against *Ichthyophthirius multifiliis*. Initial studies by Bauer (1952) showed that in carp recovering from an infection, the number of parasites able to establish in a challenge infection was reduced. According to him, resistance to re-infection persisted only for about two weeks, and then declined. Hines and Spira (1973; 1974a; 1974b; 1974c), however, have shown that carp exposed to a sublethal infection showed complete recovery and were free of the parasite after 21 days. Heavy infections of the parasite killed the fish. When the initial infection was small, a second generation of parasites could establish on the fish, but in reduced numbers. A third generation could never be established. In fish recovering from an infection, the parasites were found in unusual sites, and particularly on the fins. Fish that had recovered remained free of infection in an infective environment for 8 months, were refractory to infective doses that killed susceptible control fish, and could not act as carriers of the parasite. The response was not due to changes in the skin or to the amount of mucus produced, but rather to a qualitative change in the mucus. Both mucus and serum from immune fish immobilized free-swimming stages of the parasite, suggesting that serum factors were also involved. The response was very specific, and did not affect concurrent infections of *Trichodina* and *Dactylogyrus*. It thus appeared as if carp could exhibit true, specific, post-invasive immunity to *Ichthyophthirius*. The effects of this response on a whole population of the parasite were not examined.

It thus appears that although fish are not able to mount classic immune responses involving antibodies against their parasites, some species are able to mount responses involving mucus changes that are effective against some species of ectoparasites. These responses may be general and short-term, or specific and long-term, but in both cases confer resistance to re-infection. The responses undoubtedly serve to protect the fish, and regulate the parasites to some extent, but it has not yet been demonstrated that they can produce long term stability in a natural parasite population that is subject also to large changes in density-independent climatic factors.

REGULATION BY COMPETITION

Interspecific Competition

Interactions between two or more species of parasites have been reported on numerous occasions in fish. They may take the form of positive or negative associations between species (Noble, 1961; Noble, King, and Jacobs, 1963; Thomas, 1964) or they may involve competitive exclusion of species, niche diversification and site segregation (Chappell, 1969b; MacKenzie and Gibson, 1970; Holmes, 1971; 1973). This whole field has recently been reviewed by Crompton (1973) and Halvorsen (in press).

Whilst there is no doubt that in many cases the presence of one species of parasite in a fish may affect the distribution of another, and so indirectly its fecundity, or even affect the numbers of the other species directly (Cross, 1934; Chappell, 1969b), it is very difficult to see how competitive interactions between species can regulate, as opposed to merely affect, their infrapopulation densities. In order for one species population to regulate that of another, they must share a common regulatory mechanism, and this mechanism must respond to changes in population density

of each species separately and of both species together.
Unless it does so, a positive feedback situation is
more likely to arise. For example, if two species compete, the most successful species generally depresses
the population density of the other, and in the absence
of niche diversification and habitat segregation between them, the less successful species will disappear.
This indeed is what tends to happen when *Dactylogyrus
vastator* and *D. extensus* are present on the gills of
the same fish (Paperna, 1964). Although Schad (1966)
has suggested that the host can mediate the competition
and that this mediation may involve cross-immunity, the
general absence of effective immunological responses by
fish to their parasites suggests that such a mechanism
would be of little importance in regulating fish parasite populations. Simmons and Laurie (1972) have also
suggested that since mixed species infections of
Gyrocotyle spp. are exceptionally rare, the mechanism
that prevents over-infection by a single species may
also operate against other species. No evidence is
forthcoming to support this suggestion, however, nor do
they show how such a mechanism could regulate the densities of the other species. Although, therefore, it
is theoretically possible, in the complete absence of
any evidence for common regulatory mechanisms, it must
be concluded that interspecific competition is of no
importance in the regulation of fish parasite populations.

Intraspecific Competition

Intraspecific competition is generally accepted to
be widespread amongst parasite infrapopulations, and
especially in cestode populations where it is more frequently referred to as the crowding effect. Its typical manifestations are a decrease in the mean size of a
parasite as its population density in a host individual
increases, and a reduction in the number of eggs produced per parasite (Read, 1951; 1959; Roberts, 1961;
Jones and Tan, 1971; Hager, 1941; Hesselberg and
Andreassen, 1975; Ghazal and Avery, 1974). Parasite
longevity is also reduced on occasion. It has also
been accepted that the competition was for carbohydrate
resources (Read, 1959). Since all the effects of

crowding tend to reduce the infrapopulation size and
they are inversely related to population density, and
since intraspecific competition is widely accepted as a
possible population regulatory mechanism amongst free-
living animals, it seems reasonable to expect that it
could also be a mechanism regulating population densities of fish parasites.

More recently, however, it has become clear that
caution is needed in interpreting the population consequences of the crowding effect amongst cestodes. In
the first place there is increasing evidence that the
effect may not be due to overcrowding and competition,
but to a host immune response (Befus, 1975; Hesselberg
and Andreassen, 1975). In the second place, the relationship between parasite density and fecundity is
not a simple one and may be very complex. Although the
mean number of eggs per parasite declines, the total
number of eggs produced per infected host may increase,
remain steady or decrease (Hager, 1941; Jones and Tan,
1971; Ghazal and Avery, 1974). Finally, it is clear
that in some cases there is an optimum parasite density
per infected host, and fecundity declines above and
below this level (Boxshall, 1974a; 1974b; Halvorsen and
Andersen, 1974). Low parasite densities may reduce the
probability of parasites meeting and mating, and in
natural situations because of the overdispersion of
parasites in their intermediate hosts it would be common for definitive hosts to be infected with several
parasites at once.

Intraspecific competition, rather surprisingly,
appears to be very uncommon in fish parasites, and has
been demonstrated in two species only. Both of these
are acanthocephalans. Holmes, Hobbs, and Leong (in
this symposium) and Leong (1975) have shown that the
suprapopulation of *Metechinorhynchus salmonis* in Cold
Lake, Alberta (Canada), is regulated by intraspecific
competition within the infrapopulation in whitefish.
This results in reduced egg output, i.e. density-dependent natality at high parasite densities. Since this
situation is discussed in detail in this volume, it
will not be considered further here.

The second case of population regulation by intraspecific competition is that of *Pomphorhynchus laevis*.
This parasite is found in every species of fish in the

River Avon, although only in chub, *Leuciscus cephalus*, and barbel, *Barbus barbus*, and possibly trout, *Salmo trutta* and *S. irrideus*, does a significant part of the suprapopulation reach full sexual maturity (Hine and Kennedy, 1974a). These species can thus be regarded as its preferred hosts. For technical reasons, however, the changes in the parasite population in fish throughout the year were followed in dace, *Leuciscus leuciscus*. The limited data available suggests that the changes were similar in chub and barbel.

The parasite infrapopulation size in dace remained constant, apart from small irregular variations, throughout the year, and showed no evidence of seasonal trends in incidence or intensity of infection (Kennedy and Hine, 1974b). Both recruitment and mortality of parasites occurred throughout the year. Temperature had very little effect upon the establishment of *P. laevis* in fish; at warmer temperatures, corresponding to summer temperatures in the river, the establishment rate of the parasites in fish was reduced slightly and so mortality increased, but this was compensated for by the increased feeding intensity, and so, increased recruitment rate of parasites by the fish (Kennedy, 1972a). The parasite infrapopulation thus appeared to be in a state of dynamic equilibrium throughout the year, such that recruitment into the population and mortality were continuous, equal, and opposed, and infection levels remained steady. The population levels in any individual fish appeared to depend entirely upon the immigration rate of parasites into that fish, and thus upon its diet. Nevertheless, the constancy of the infrapopulation level suggested that regulation was taking place. This was confirmed by a long-term study over nine years, which showed that the infrapopulation levels of the parasite in both dace and in its intermediate host, *Gammarus pulex* (the only species to serve as intermediate host in this locality), remained constant over the whole period, and that variations within any one year were greater than variations between years (Kennedy and Rumpus, in press).

This regulation was not, however, due to density-dependent mortality. The loss of parasites from fish was shown experimentally to be independent of the size of the infection and to bear no relationship to the number of parasites already present in the fish (Kennedy, 1974), and indeed it appeared that *P. laevis* could continue to establish in fish as long as there was physically room for it. No interaction between *P. laevis* and any other species of parasite in the same fish could be detected (Kennedy, in press), nor was there any evidence of seasonal niche segregation by the parasite (Kennedy, 1972b). Apart from local damage to the fish intestine, *P. laevis* did not cause any harm to its host and certainly did not induce host mortality (Kennedy and Hine, 1974a). Fish were able to respond to the parasite by producing antibodies, but these were only produced in the preferred host since the antigen appeared to be an excretory/secretory product of the sexually mature parasite, and they had no effect upon the parasite (Harris, 1972). No density-dependent parasite mortality was observed in the intermediate host either, nor did *P. laevis* cause mortality of *G. pulex* (Rumpus, 1974).

Since no density-dependent mortality of *P. laevis* could be detected, regulation had to be due to density-dependent natality. This could only occur in chub and barbel, the preferred hosts. *P. laevis* does show a preference for a particular region of the alimentary tract of fish, but it can establish and grow throughout the entire length of the alimentary tract (Kennedy, Broughton, and Hine, 1976). It establishes in the first vacant and suitable site, and thereafter does not move. In heavy infections in an individual fish, the number of parasites living outside of the preferred site is greater than in light ones, and these parasites are smaller than those in the preferred site. Thus in crowded populations, the mean size of the parasites is smaller, the number of ovarian balls produced is less, and egg production decreases (Kennedy, unpublished observations). Overcrowding may therefore result in competition for space and, subsequently, probably in decreased natality, which is hence density-dependent. Regulation of the *P. laevis* population in the River Avon shows close similarities to the way in which the suprapopulation of *Metechinorhynchus salmonis* is regulated in Cold Lake (see Holmes, Hobbs, and Leong, this volume); both species populations are regulated by density-dependent natality, and the suprapopulations in both cases are regulated by a single feedback control mechanism operating in a single host species. Since, however, these are the only two examples known, it must be concluded that on present evidence, regulation of fish parasite populations by intraspecific competition is rare and exceptional.

REGULATION BY OTHER MECHANISMS

In a few instances, evidence for regulation of infrapopulations of parasites in fish has been claimed, but no real indication of the mechanism responsible for it has been observed. Williams and Halvorsen (1971) noted that 70% of *Gadus morhua*, infected with the tapeworm, *Abothrium gadi*, contained only a single parasite. Infections with more than one cestode were more common in older fish. The parasite showed a marked preference for a particular site in the alimentary canal. The authors believed, though without any proof, that a number of larvae were probably ingested at the same time, but that only one survived. They did not believe that crowding was responsible for this, but attributed it to an intrinsic factor that controlled numbers and prevented crowding. It is possible, however, that the high incidence of single parasite infections could merely be a reflection of low levels of overdispersion, the infrequency of the parasite in its intermediate host and the overall low parasite density. In any event, the evidence for regulation of *Abothrium gadi* is entirely circumstantial and very slight at that.

A more convincing case is that of species of *Gyrocotyle* in their chimaerid hosts. Halvorsen and Williams (1968) observed that the majority of infections in *Chimaera monstrosa* consisted of only two sexually mature parasites per fish. Fish did not become infected at all until they were 20 cm long, when they started to feed. Some of these young fish contained large numbers of parasite post-larvae, of the same age, although most still contained only one or two parasites. The authors believed, therefore, that infections in older fish did not arise from the acquisition of one parasite and then later another, but that only two parasites survived from a multiple infection as a consequence of a regulatory mechanism that ensured that only a single sexual unit of one male and one female parasite survived in any individual fish. The only suggestion they made with respect to the mechanism of this regulation was conditioning of the habitat in some way. Simmons and Laurie (1972), studying different species of *Gyrocotyle* in different host species, confirmed that the majority of infections consisted of only two sexually mature parasites per fish. Second in frequency were infections with only one parasite. They found, however, that small fish were characterized by only single parasite infections and that there was no evidence for massive infections with several postlarvae. In addition, a few mature fish were found to contain multiple infections. They were impressed also by the low frequency of infections with more than one species of *Gyrocotyle*, and argued that a mechanism existed for preventing mixed species infections and for supressing supra-infection by a single species. This mechanism, they suggested, was akin to the crowding phenomenon; it might involve chemical mediation of the habitat, although infections with two parasites did not arise from selection from a multiple infection but from the accumulation of parasites with time as a consequence of repeated infections by single parasites and the loss of parasites when two adults were already present. Such an explanation would also fit the data of Halvorsen and Williams (1971). Although, therefore, it appears likely that *Gyrocotyle* populations are regulated by intraspecific competition, until this is firmly established and the mechanism clearly demonstrated, it seems advisable to consider them under a separate heading.

A particularly interesting example of long-term stability and infrapopulation regulation by unknown mechanisms is that of *Eubothrium salvelini* in smolts of *Onchorhynchus nerka* in Babine Lake, Canada, (Smith, 1973). The incidence of infection in smolt runs in 15 years, over a 20 year period, remained fairly steady (mean 32%, range 25%-45%), except for two short periods (1955-1957 and 1970-1971) when it dropped to 12% (range 6%-18%). Daily infection rates during the smolt run were very variable, but reached a maximum in the last few days and the incidence was inversely proportional to the number of migrants. During the run, the mean weight of parasites per fish increased until the end of May, declined sharply in June, then stabilized at a lower level. In pre-smolts (those not migrating that year), the incidence of parasites increased to a maximum at the end of July, then declined and stabilized at a lower level. The stable incidence level of 30% persisted until the fish emigrated as smolts.

Smith believed this stability in pre-smolts was due to the fact that copepods, the intermediate host, reached a maximum number in the lake in spring and a nadir after July. Salmon no longer fed on them after this time, so new infections could not be acquired, and since the parasite could survive up to 24 months in fish, there was little or no mortality in the infrapopulation. The inverse relationship between incidence of infection and number of smolts migrating was believed to be due to bursts of healthy fish being superimposed upon a more regular output of infected ones from the lake. These latter migrated more slowly, due to the effect of the parasite, and so constituted the bulk of the fish in the last few days of the run. The longer the infected fish took in transit through the lake, the greater the probability that they would shed eggs before leaving it. If copepod levels were low, or the run early and eggs shed before copepods appeared, a low incidence of infection would ensue in the following year. This was the explanation for the two periods of lower than average incidence during the runs. The seasonal pattern of intensity changes during the run remained constant. The decrease in intensity after May was held to be due to a diminution of parasite numbers in some or all heavily infected fish. The mechanism of this was not known; since there was no immigration into the parasite infrapopulation it could not have involved a dynamic equilibrium or reduced rates of establishment in crowded infections. Competition was also considered unlikely, but there appeared to be some form of density-dependent mortality of parasites. In addition, according to Smith, parasite-induced, host mortality also occurred. Heavily infected fish were smaller (Dombrowski, 1955), fatigued more readily (Smith and Margolis, 1970) and were more likely to die of fatigue, were more susceptible to death by predation and to death during the passage to the sea as a consequence of the osmotic stress and energy demands on them. Although Smith postulated these two causes of mortality, it seems possible to this author that the death of heavily infected fish alone could account for the diminution of parasite numbers and decline in numbers of heavily infected fish after May. Whatever the mechanism, and this clearly needs further investigation, the population appeared to be regulated and stable due to the operation of density-dependent factors, although density-independent factors had a very pronounced influence upon the levels of infection.

There is no evidence at all as yet to suggest that retardation of development occurs amongst fish nematodes. This phenomenon is widespread amongst nematode parasites of birds and mammals, although there is controversy over whether it is immunologically or climatically induced (see the contribution of Schad to this volume). If it is an immunological phenomenon, its absence in fish nematodes is not surprising in view of the comments made earlier about fish immune systems. If it is a climatic phenomenon, the explanation may lie in the fact that mammalian nematodes showing retardation all have direct life cycles and are capable of inducing host antibody responses, whereas all fish nematodes appear to have indirect life cycles and do not appear to be immunogenic.

CONCLUSIONS

Despite the large number of studies that have been carried out upon the changes in infrapopulations of parasites in fish, far too little of the information obtained is of the right quality to provide useful data about the nature and effectiveness of parasite regulatory mechanisms. In the majority of studies, no density-dependent factors have been identified at all, and only density-independent factors have been shown to affect the parasite populations. Of these, the most widespread and important appear to be water temperature and fish behavior, both dietary and social. Thus, on the evidence available at present, it must be concluded that the majority of fish parasite populations are unregulated and hence unstable. Where water temperature shows a regular oscillatroy cycle within reasonably constant limits and the fish feeding behavior remains fairly constant, the parasite infrapopulation size may also oscillate regularly and show similar changes in different localities. It may even persist for several years, but despite this appearance of stability, the population is still unstable and liable to extinction as a consequence of unusual changes in climate.

In only a few studies have density-dependent factors, capable of regulating infrapopulations of parasites in fish, been identified at all. In fewer still, has it been shown that they effectively regulate the population, or indeed that the population is actually regulated. Only the studies of *Pomphorhynchus laevis* and *Eubothrium salvelini* have continued for a sufficient length of time to make it reasonably certain that the parasite population is regulated. With so little information it thus becomes almost impossible to make any general comments upon regulatory mechanisms or their occurrence, or correlate them with the type of life cycle or host. Nevertheless, a few significant points do emerge. These suggest that the type of mechanism and the point at which it occurs in the life cycle of the parasite are such as to confer the maximum ecological advantage for the parasite. Thus, regulation by parasite-induced, host mortality appears only to occur when the fish is the intermediate host of the parasite. Since the induction of such mortality almost invariably renders the fish more susceptible to predation by the parasite's definitive host, this has a clear ecological advantage to the parasite. For ectoparasites with a direct life cycle, the most advantageous, and indeed virtually the only, place for regulation to occur is when they are living on the fish host, and this is the only group to show regulation involving an effective fish response. For endoparasites employing fish as their definitive host, the only regulatory mechanism to be positively identified at all is intraspecific competition, and then it is of rare occurrence. Since the population levels of both *P. laevis* and *M. salmonis* were very high in the localities studied, it may be that these were unusual situations, and that competition only operates as a last resort and in very overcrowded populations. Regulation involving host immune responses is clearly very rare indeed in fish, if it even occurs at all, and this is entirely consistent with the evolutionary state of development of the fish immune response.

The failure to demonstrate the existence of regulatory mechanisms in most instances may be the result of the limited nature of the studies. It could be that regulation often occurs in the infrapopulation in the intermediate host; a regulatory mechanism at any stage in the life cycle is still sufficient to regulate the whole suprapopulation, and few investigations have studied the parasite in all its hosts simultaneously. It may also be true, however, that most fish parasite populations are unstable and unregulated. Even when regulatory mechanisms have been identified, density-independent factors still exert pronounced effects upon the parasite population and may result in its extinction (populations of *Schistocephalus*); and long term oscillations, despite being predicted by models, have not been demonstrated. It may indeed be that the conditions, predicted by the models, for instability are far more widespread than those for stability, and that, as is the case so often with free-living animals, population levels are depressed for most of the time by density-independent factors and kept below the point at which density-dependent factors operate. Fish parasite populations nevertheless, like those of free-living animals, do have regulatory mechanisms that can operate at high densities. What is needed now are many more studies over a longer period and in greater depth and breadth to assist our understanding of how and when they operate.

ACKNOWLEDGEMENTS

I should like to thank the University of Tromso for providing me with the facilities and opportunity to complete this review. I should also like to thank the many people with whom I have discussed the many ideas and conclusions presented here, and in particular Prof. G. W. Esch, for ensuring my presence at this meeting; Prof. J. Holmes, for stimulating me to do some further thinking; and Prof. O. Halvorsen, for giving me the opportunity to do so and then for his comments upon the result of it. Finally, I should like to thank the Royal Society and the University of Exeter for the financial support for my visit to New Orleans.

LITERATURE CITED

ANDERSON, R. M. 1974a. Mathematical models of host-helminth interactions. *In:* M. B. Usher and M. H. Williamson (eds.), Ecological Stability. Chapman and Hall, London.

ANDERSON, R. M. 1974b. An analysis of the influence of host morphometric features upon the population dynamics of *Diplozoon paradoxum* (Nordman, 1832). J. Anim. Ecol. 43: 873-887.

ANDERSON, R. M. 1974c. Population dynamics of the cestode *Caryophyllaeus laticeps* (Pallas, 1781) in the bream (*Abramis brama* L.). J. Anim. Ecol. 43: 305-321.

ANDERSON, R. M. In press. Dynamic aspects of parasite population ecology. *In:* C. R. Kennedy (ed.), Ecological Aspects of Parasitology. Elsevier Press, Amsterdam.

ANDERSON, R. M. and P. J. WHITFIELD. 1975. Survival characteristics of the free-living cercarial population of the ectoparasite digenean *Transversotrema patialensis* (Soporker, 1924). Parasitology 70: 295-310.

ARME, C. 1968. Effects of the plerocercoid larva of a pseudophyllidean cestode, *Ligula intestinalis*, on the pituitary gland and gonads of its host. Biol. Bull. 134: 15-25.

ARME, C. and R. W. OWEN. 1967. Infections of the three-spined stickleback, *Gasterosteus aculeatus* L., with special reference to pathological effects. Parasitology 57: 301-314.

ARME, C. and R. W. OWEN. 1968. Occurrence and pathology of *Ligula intestinalis* infections in British fishes. J. Parasitol. 54: 272-280.

ARME, C. and M. WALKEY. 1970. The physiology of fish parasites. *In:* A. E. R. Taylor and R. Muller (eds.), Aspects of Fish Parasitology (Vol. 8). Symposium Brit. Soc. Parasitol. Blackwell Sci. Pub., Oxford and Edinburgh.

BAUER, O. N. 1952. Immunity of fish occurring in infections with *Ichthyophthirius multifiliis* (Foquet). *Dokl. Acad. Nauk SSSR 93:* 377-379.

BAUER, O. N. 1961. Parasitic diseases of cultured fishes and methods of their prevention and treatment. *In:* V. A. Dogiel, G. K. Petrushevski and Yu. I. Polyanski (eds.), Parasitology of Fishes. Oliver and Boyd, London.

BAUER, O. N. 1962. The ecology of parasites of freshwater fish. *In:* Parasites of Freshwater Fish. and the Biological Basis for Their Control. IPST, Jerusalem.

BAUER, O. N., V. A. MUSSELIUS, and YU. A. SRELKOV. 1973. Diseases of Pond Fishes. IPST, Jerusalem.

BECKER, C. D. and W. D. BRUNSON. 1967. *Diphyllobothrium* (Cestoda) infections in salmonids from three Washington lakes. *J. Wildl. Mgmt. 31:* 813-824.

BEFUS, A. D. 1975. Secondary infections of *Hymenolepis diminuta* in mice: effect of varying worm burdens in primary and secondary infections. *Parasitology 71:* 61-75.

BETTERTON, C. 1974. Studies on the host specificity of the eye fluke, *Diplostomum spathaceum*, in brown and rainbow trout. *Parasitology 69:* 11-29.

BORGSTROM, R. 1970. Studies of the helminth fauna of Norway XVI: *Triaenophorus nodulosus* (Pallas, 1760) (Cestoda) in Bogstad Lake. III. Occurrence in pike *Esox lucius* L. *Nor. J. Zool. 18:* 209-216.

BOXSHALL, G. A. 1974a. The population dynamics of *Lepeophtheirus pectoralis* (Muller): seasonal variation in abundance and age structure. *Parasitology 69:* 361-371.

BOXSHALL, G. A. 1974b. The population dynamics of *Lepeophtheirus pectoralis* (Muller): dispersion pattern. *Parasitology 69:* 373-390.

BRADLEY, D. J. 1974. Stability in host-parasite systems. *In:* M. B. Usher and M. H. Williamson (eds.), Ecological Stability. Chapman and Hall, London.

CHAPPELL, L. H. 1969a. The parasites of the three-spined stickleback *Gasterosteus aculeatus* from a Yorkshire Pond. 1. Seasonal variation of parasite fauna. *J. Fish. Biol. 1:* 137-152.

CHAPPELL, L. H. 1969b. Competitive exclusion between

two intestinal parasites of the three-spined stickle-back, *Gasterosteus aculeatus* L. *J. Parasitol. 55:* 775-778.

CHUBB, J. C. 1963. Seasonal occurrence and maturation of *Triaenophorus nodulosus* (Pallas, 1781) (Cestoda: Pseudophyllidea) in the pike *Esox lucius* L. of Llyn Tegid. *Parasitology 53:* 419-433.

CHUBB, J. C. 1964. Observations on the occurrence of the plerocercoids of *Triaenophorus nodulosus* (Pallas, 1781) (Cestoda: Pseudophyllidea) in the perch *Perca fluviatilis* L. of Llyn Tegid (Bala Lake), Merioneth. *Parasitology 54:* 481-491.

CROFTON, H. D. 1971. A model of host-parasite relationships. *Parasitology 63:* 343-364.

CROMPTON, D. W. T. 1973. The sites occupied by some parasitic helminths in the alimentary tract of vertebrates. *Biol. Rev. 48:* 27-83.

CROSS, S. X. 1934. A probable case of non-specific immunity between two parasites of ciscoes of the Trout Lake region of Northern Wisconsin. *J. Parasitol. 20:* 244-245.

DAVIES, R. B., W. T. BURKHARD, and C. P. HIBLER. 1973. Diplostomosis in North Park, Colorado. *J. Wildl. Dis. 9:* 362-367.

DENCE, W. A. 1958. Studies on *Ligula*-infected common shiners (*Notropis cornutus* Agassiz) in the Adirondacks. *J. Parasitol. 44:* 334-338.

DOBBEN, W. H. VAN. 1952. The food of the cormorant in the Netherlands. *Ardea 40:* 1-63.

DOMBROWSKI, E. 1955. Cestode and nematode infections of sockeye smolts from Babine Lake, British Columbia. *J. Fish. Res. Bd. Can. 12:* 93-96.

ESCH, G. W., J. W. GIBBONS, and J. E. BOURQUE. 1975. An analysis of the relationship between stress and parasitism. *Amer. Mid. Nat. 93:* 339-353.

ESCH, G. W., W. C. JOHNSON, and J. R. COGGINS. 1975. Studies on the population biology of *Proteocephalus ambloplitis* (Cestoda) in the smallmouth bass. *Proc. Okla. Acad. Sci. 55:* 122-127.

EURE, H. E. and G. W. ESCH. 1974. Effects of thermal effluent on the population dynamics of helminths in largemouth bass. *In:* J. W. Gibbons and R. R. Sharitz (eds.), Thermal Ecology. AEC Symposium Series (CONF. 730505).

FISCHER, H. 1967. The life cycle of *Proteocephalus fluviatilis* (Bangham) (Cestoda) from smallmouth bass, *Micropterus dolomieui* Lacepede. *Canad. J. Zool. 46:* 569-579.

FISCHER, H. and R. S. FREEMAN. 1969. Penetration of parenteral plerocercoids of *Proteocephalus ambloplitis* (Leidy) into the gut of smallmouth bass. *J. Parasitol. 55:* 766-774.

FLETCHER, T. C. and P. T. GRANT. 1969. Immunoglobulins in the serum and mucus of plaice, *Pleuronectes platessa. Biochem. J. 115:* 65.

FRASER, P. G. 1960. The occurrence of *Diphyllobothrium* in trout, with special reference to an outbreak in the West of England. *J. Helminthol. 34:* 59-72.

GHAZAL, A. M. and R. A. AVERY. 1974. Population dynamics of *Hymenolepis nana* in mice: fecundity and the crowding effect. *Parasitology 69:* 403-415.

HAGER, A. 1941. Effects of dietary modification of host rats on the tapeworm *Hymenolepis diminuta. Iowa State Coll. J. of Sci. 15:* 127-153.

HALVORSEN, O. 1969. Studies on the helminth fauna of Norway XIII. *Diplozoon paradoxum* Nordman, 1832, from roach, *Rutilus rutilus* (L.), bream *Abramis brama* (L.), and hybrid of roach and bream. Its morphological adaptability and host specificity. *Nytt. Mag. Zool. 17:* 93-103.

HALVORSEN, O. 1970. Studies of the helminth fauna of Norway XV. On the taxonomy and biology of plerocercoids of *Diphyllobothrium* Cobbold, 1858 (Cestoda: Pseudophyllidea) from N. W. Europe. *Nor. J. Zool. 18:* 113-174.

HALVORSEN, O. 1971. Studies on the helminth fauna of Norway XIX. The seasonal cycle and microhabitat preference of the leech *Cystobranchus mammillatus* (Malm 1863) parasitising burbot, *Lota lota* (L.). *Nor. J. Zool. 19:* 177-180.

HALVORSEN, O. In press. Negative interactions among parasites. *In:* C. R. Kennedy (ed.), Ecological Aspects of Parasitology. Elsevier Press, Amsterdam.

HALVORSEN, O. and K. ANDERSEN. 1974. Some effects of population density in infections of *Diphyllobothrium dendriticum* (Nitzsch) in golden hamster

(*Mesocricetus auratus* Waterhouse) and common gull
(*Larus canus* L.). *Parasitology* 69: 149-160.
HALVORSEN, O. and H. H. WILLIAMS. 1968. Studies on
the helminth fauna of Norway IX. *Gyrocotyle*
(Platyhelminths) in *Chimaera monstrosa* from Oslo
Fjord, with emphasis on its mode of attachment and
regulation in the degree of infection. *Nytt. Mag.
Zool.* 15: 130-142.
HARRIS, J. E. 1972. The immune response of a cyprinid
fish to infections of the acanthocephalan *Pomphorhynchus laevis*. *Internat. J. Parasitol.* 2: 459-469.
HARRIS, J. E. 1973a. The apparent inability of cyprinid fish to produce skin-sensitizing antibody. *J.
Fish Biol.* 5: 535-540.
HARRIS, J. E. 1973b. The immune responses of dace
Leuciscus leuciscus (L.) to injected antigenic
materials. *J. Fish Biol.* 5: 261-276.
HESSELBERG, C. A. and J. ANDREASSEN. 1975. Some influences of population density on *Hymenolepis
diminuta* in rats. *Parasitology* 71: 517-523.
HICKEY, M. D. and J. R. HARRIS. 1947. Progress of the
Diphyllobothrium epizootic at Poulaphouca Reservoir, Co. Wicklow, Ireland. *J. Helminthol.* 22:
13-28.
HINE, P. M. and C. R. KENNEDY. 1974a. Observations on
the distribution, specificity and pathogenicity of
the acanthocephalan *Pomphorhynchus laevis*,
(Muller). *J. Fish Biol.* 6: 521-535.
HINE, P. M. and C. R. KENNEDY. 1974b. The population
biology of the acanthocephalan *Pomphorhynchus
laevis* (Muller) in the River Avon. *J. Fish Biol.*
6: 665-679.
HINES, R. S. and D. T. SPIRA. 1973. *Ichthyophthirius
multifiliis* (Fouquet) in the mirror carp, *Cyprinus
carpio* L. I. Course of infection. *J. Fish Biol.*
5: 385-392.
HINES, R. S. and D. T. SPIRA. 1974a. *Ichthyophthirius
multifiliis* (Fouquet) in the mirror carp, *Cyprinus
carpio* L. II. Pathology. *J. Fish Biol.* 6: 189-196.
HINES, R. S. and D. T. SPIRA. 1974b. *Ichthyophthirius
multifiliis* (Fouquet) in the mirror carp, *Cyprinus
carpio* L. IV. Physiological dysfunction. *J.*

Fish Biol. 6: 365-372.

HINES, R. S. and D. T. SPIRA. 1974c. *Ichthyophthirius multifiliis* (Fouquet) in the mirror carp, *Cyprinus carpio* L. V. Acquired immunity. *J. Fish Biol. 6:* 373-378.

HOFFMAN, G. L. and C. E. DUNBAR. 1961. Mortality of Eastern brook trout caused by plerocercoids (Cestoda: Pseudophyllidea) (Diphyllobothriidae) in the heart and viscera. *J. Parasitol. 47:* 399-400.

HOLMES, J. C. 1971. Habitat segregation in sanguinicolid blood flukes (Digenea) of scorpaenid rockfishes (Perciformes) on the Pacific Coast of North America. *J. Fish. Res. Bd. Can. 28:* 903-909.

HOLMES, J. C. 1973. Site selection by parasitic helminths: interspecific interactions, site segregation, and their importance to the development of helminth communities. *Can. J. Zool. 51:* 333-347.

HOLMES, J. C. and W. M. BETHEL. 1972. Modification of intermediate host behavior by parasites. *In:* E. U. Canning and C. A. Wright (eds.), Behavioral Aspects of Parasite Transmission. *Zoo. J. Linn. Soc. 51 (Suppl. 1):* 123-149.

HOPKINS, C. A. 1959. Seasonal variations in the incidence and development of the cestode *Proteocephalus filicollis* (Rud. 1810) in *Gasterosteus aculeatus* (L. 1766). *Parasitology 49:* 529-542.

JAHN, T. L. and L. R. KUHN. 1932. The life history of *Epibdella melleni* Mac Callum, 1927, a monogenetic trematode parasitic on marine fishes. *Biol. Bull. 62:* 89-111.

JONES, A. W. and B. D. TAN. 1971. Effect of crowding upon growth and fecundity in the mouse bile duct tapeworm *Hymenolepis microstoma*. *J. Parasitol. 57:* 88-93.

KENNEDY, C. R. 1968. Population biology of the cestode *Caryophyllaeus laticeps* (Pallas, 1781) in dace, *Leucuscus leuciscus* L. of the River Avon. *J. Parasitol. 54:* 538-543.

KENNEDY, C. R. 1969. Seasonal incidence and development of the cestode *Caryophyllaeus laticeps* (Pallas) in the River Avon. *Parasitology 59:* 783-794.

KENNEDY, C. R. 1970. The population biology of helminths of British freshwater fish. *In:* A. E. R.

Taylor and R. Muller (eds.), Aspects of Fish Parasitology (Vol. 8). *Symposium Brit. Soc. Parasitol.*, Blackwell Sci. Pub., Oxford and Edinburgh.

KENNEDY, C. R. 1971. The effects of temperature upon the establishment and survival of the cestode *Caryophyllaeus laticeps* in orfe, *Leuciscus idus*. *Parasitology 63:* 59-66.

KENNEDY, C. R. 1972a. The effect of temperature and other factors upon the establishment and survival of *Pomphorhynchus laevis* in goldfish, *Carrasius auratus*. *Parasitology 65:* 283-294.

KENNEDY, C. R. 1972b. Parasite communities in freshwater ecosystems. *In:* R. B. Clark and R. J. Wootton (eds.), Essays in Hydrobiology. University Press, Exeter.

KENNEDY, C. R. 1974. The importance of parasite mortality in regulating the population size of the acanthocephalan *Pomphorhynchus laevis* in goldfish. *Parasitology 68:* 93-101.

KENNEDY, C. R. 1975. Ecological Animal Parasitology. Blackwell Sci. Pub., Oxford and Edinburgh.

KENNEDY, C. R. In press. Reproduction and dispersal. *In:* C. R. Kennedy (ed.), Ecological Aspects of Parasitology. Elsevier Press, Amsterdam.

KENNEDY, C. R. and P. M. HINE. 1969. Population biology of the cestode *Proteocephalus torulosus* (Batsch) in dace *Leuciscus* (L.) of the River Avon. *J. Fish Biol. 1:* 209-219.

KENNEDY, C. R. and S. F. LIE. 1976. The distribution and pathogenicity of larvae of *Eustrongylides* (Nematoda) in brown trout *Salmo trutta* L. in Fernworthy Reservoir, Devon. *J. Fish Biol. 8:* 293-302.

KENNEDY, C. R. and A. RUMPUS. In press. Long term changes in the size of the *Pomphorhynchus laevis* (Acanthocephala) population in the River Avon. *J. Fish Biol.*

KENNEDY, C. R. and P. J. WALKER. 1969. Evidence for an immune response by dace, *Leuciscus leuciscus*, to infections by the cestode *Caryophyllaeus laticeps*. *J. Parasitol. 55:* 579-582.

KENNEDY, C. R., P. F. BROUGHTON, and P. M. HINE. 1976. The sites occupied by the acanthocephalan

Pomphorhynchus laevis in the alimentary canal of fish. *Parasitology 72:* 195-206.

KERR, T. 1948. The pituitary in normal and parasitised roach (*Leuciscus rutilus* Flem.). *Q. J. Mic. Soc. 89:* 129-137.

KHALIL, L. F. 1964. On the biology of *Macrogyrodactylus polypteri*, Malmberg 1956, a monogenetic trematode on *Polypterus senegalus* in the Sudan. *J. Helminthol. 38:* 219-222.

KUPERMAN, B. I. 1973. Infection of young perch by the tapeworm *Triaenophorus nodulosus*. *Verh. Internat. Verein. Limnol. 18:* 1697-1704.

LESTER, R. G. 1971. The influence of *Schistocephalus* plerocercoids on the respiration of *Gasterosteus* and a possible resulting effect on the behavior of the fish. *Can. J. Zool. 49:* 361-366.

LESTER, R. J. G. 1972. Attachment of *Gyrodactylus* to *Gasterosteus* and host response. *J. Parasitol. 58:* 717-722.

LESTER, R. J. G. and J. R. ADAMS. 1974a. *Gyrodactylus alexandrei:* reproduction, mortality and effect on its host *Gasterosteus aculeatus*. *Can. J. Zool. 52:* 827-833.

LESTER, R. J. G. and J. R. ADAMS. 1974b. A simple model of a *Gyrodactylus* population. *Internat. J. Parasitol. 4:* 497-506.

LEONG, R. 1975. The parasites of fishes from Cold Lake, Alberta. Ph.D. thesis: University of Alberta.

LIEN, L. 1970. Studies of the helminth fauna of Norway. XIV. *Triaenophorus nodulosus* (Pallas, 1760) (Cestoda) in Bogstad Lake. II. Development and life span of the plerocercoids in perch (*Perca fluviatilis* L., 1758). *Nor. J. Zool. 18:* 85-96.

LLEWELLYN, J. 1962. The life histories and population dynamics of monogenean gill parasites of *Trachurus trachurus* (L.). *J. Mar. Biol. Ass. U. K. 42:* 587-600.

LOCKE, L., J. DE WITT, C. MENZIE, and J. KERWIN. 1964. A merganser dieoff associated with larval *Eustrongylides*. *Avian Dis. 8:* 420-427.

LOPUKHINA, A. M., YU. A. STRELKOV, N. B. CHERNYSHOVA, and O. N. YUNCHIS. 1973. The effects of parasites on the abundance of young fish in natural water bodies. *Verh. Internat. Verein. Limnol.* 18: 1705-1712.

MACKENZIE, K. and D. GIBSON. 1970. Ecological studies of some parasites of plaice, *Pleuronectes platessa* (L.), and flounder, *Platichthys flesus* (L.). In: A. E. R. Taylor and R. Muller (eds.), Aspects of Fish Parasitology (Vol. 8). Blackwell Sci. Pub., Oxford and Edinburgh.

MACVICAR, A. M. and T. C. FLETCHER. 1970. Serum factors in *Raia radiata* toxic to *Acanthobothrium quadripartitum* (Cestoda: Tetraphyllidea), a parasite specific to *R. naevus. Parasitology* 61: 55-63.

MAY, R. M. 1973. Stability and Complexity in Model Ecosystems. Princeton University Press, Princeton.

MAY, R. M., G. R. CONWAY, M. P. HASSELL, and T. R. E. SOUTHWOOD. 1974. Time delays, density-dependent and single species oscillations. *J. Anim. Ecol.* 43: 747-770.

MEAKINS, R. H. 1974. The bioenergetics of the *Gasterosteus/Schistocephalus* host-parasite system. *Polsk. Arch. Hydrobiol.* 21: 455-466.

MILBRINK, G. 1975. Population biology of the cestode *Caryophyllaeus laticeps* in bream, *Abramis brama*, and the feeding of fish on oligochaetes. *Rep. Inst. F. W. Res. Drottning.* 54: 36-51.

MILLER, R. B. 1943. Studies on cestodes of the genus *Triaenophorus* from fish of Lesser Slace Lake, Alberta. I. Introduction and the life cycle of *Triaenophorus crassus* Forel and *T. nodulosus* (Pallas) in the definitive host, *Esox lucius. Can. J. Res. D.* 21: 160-170.

MOLNAR, K. and I. BERCZI. 1965. Demonstration of parasite specific antibodies in fish blood by agar gel diffusion precipitation test. *Z. Immun. Allergie-Forsch.* 129: 263.

MUZZALL, P. M. and F. C. RABALAIS. 1975. Studies on *Acanthocephalus jacksoni* Bullock, 1962 (Acanthocephala: Echinorhynchidae). I. Seasonal periodicity and new host records. *Proc. Helm. Soc. Wash.* 42: 31-34.

NIGRELLI, R. F. 1935. On the effects of fish mucus on *Epibdella melleni*, a monogenetic trematode of marine fishes. *J. Parasitol. 21:* 6-15.

NIGRELLI, R. F. 1937. Further studies on the susceptibility and acquired immunity of marine fishes to *Epibdella melleni*, a monogenetic trematode. *Zoologia, N. Y. 22:* 185-197.

NOBLE, E. R. 1961. The relations between *Trichodina* and metazoan parasites on gills of a fish. *Proc. First Internat. Cong. Protozool.* 521-523.

NOBLE, E. R., R. E. KING, and B. L. JACOBS. 1963. Ecology of the gill parasites of *Gillichthys mirabilis* Cooper. *Ecology 44:* 295-305.

O'ROURKE, F. J. 1961. Presence of blood antigens in fish mucus and its possible parasitological significance. *Nature, Lond. 189:* 943.

ORR, T. S. C. 1966. Spawning behavior of rudd, *Scardinius erythrophthalmus*, infected with plerocercoids of *Ligula intestinalis*. *Nature, Lond. 212:* 736.

ORR, T. S. C., C. A. HOPKINS, and G. H. CHARLES. 1969. Host specificity and rejection of *Schistocephalus solidus*. *Parasitology 59:* 683-690.

PALING, J. E. 1965. The population dynamics of the monogenean gill parasite *Discocotyle sagittata* Leuckart on Windermere trout, *Salmo trutta* L. *Parasitology 55:* 667-694.

PAPERNA, I. 1963. Dynamics of *Dactylogyrus vastator* Nybelin (Monogenea) populations on the gills of carp fry in fish ponds. *Bannidgeh. Bull. Fish. Culture, Israel 15:* 31-50.

PAPERNA, I. 1964. Competitive exclusions of *Dactylogyrus extensus* by *Dactylogyrus vastator* (Trematoda: Monogenea) on the gills of reared carp. *J. Parasitol. 50:* 94-98.

PENNYCUICK, L. 1971a. Differences in the parasite infections in three-spined sticklebacks (*Gasterosteus aculeatus* L.) of different sex, age and size. *Parasitology 63:* 407-418.

PENNYCUICK, L. 1971b. Seasonal variations in the parasite infections in a population of three-spined sticklebacks, *Gasterosteus aculeatus* L. *Parasitology 63:* 373-388.

PENNYCUICK, L. 1971c. Frequency distributions of parasites in a population of three-spined sticklebacks, *Gasterosteus aculeatus* L., with particular reference to the negative binomial distribution. *Parasitology 63:* 389-406.

PENNYCUICK, L. 1971d. Quantitative effects of three species of parasites on a population of three-spined sticklebacks, *Gasterosteus aculeatus* L. *J. Zool., Lond. 165:* 143-162.

PETRUSHEVSKI, G. K. and S. S. SHULMAN. 1961. The parasitic diseases of fishes in the natural waters of the U. S. S. R. *In:* V. A. Dogiel, G. K. Petrushevski, and Yu. I. Polyanski (eds.), Parasitology of Fishes. Oliver and Boyd, London.

POWELL, A. M. and J. C. CHUBB. 1966. A decline in the occurrence of *Diphyllobothrium* plerocercoids in the trout *Salmo trutta* L. of Llyn Padarn, Caernarvonshire. *Nature, Lond. 211:* 439.

PUTZ, R. E. and G. L. HOFFMAN. 1964. Studies on *Dactylogyrus corporalis* n. sp. (Trematoda: Monogenea) from the fall fish *Semotilus corporalis*. *Proc. Helm. Soc. Wash. 31:* 139-143.

REES, G. 1967. Pathogenesis of adult cestodes. *Helm. Abs. 36:* 1-23.

READ, C. P. 1951. The 'crowding effect' in tapeworm infections. *J. Parasitol. 37:* 174-178.

READ, C. P. 1959. The role of carbohydrates in the biology of cestodes. VIII. Some conclusions and hypotheses. *Expt. Parasitol. 8:* 365-382.

ROBERTS, L. S. 1961. The influence of population density on patterns and physiology of growth in *Hymenolepis diminuta* (Cestoda: Cyclophyllidea) in the definitive host. *Expt. Parasitol. 11:* 332-371.

RUMPUS, A. 1974. The parasites of *Gammarus pulex* in the River Avon, Hampshire. Ph.D. thesis, University of Exeter.

SCHAD, G. A. 1966. Immunity, competition and natural regulation of helminth populations. *Amer. Nat. 100:* 359-364.

SIMMONS, J. E. and J. S. LAURIE. 1972. A study of *Gyrocotyle* in the San Juan Archipelago, Puget Sound, U.S.A. with observations on the host *Hydrolagus colliei* (Lay and Bennett). *Internat. J. Parasitol. 2:* 59-77.

SMITH, H. D. 1973. Observations on the cestode *Eubothrium salvelini* in juvenile sockeye salmon *(Onchorhynchus nerka)* at Babine Lake, British

Columbia. *J. Fish. Res. Bd. Can. 30:* 947-964.
SMITH, H. D. and L. MARGOLIS. 1970. Some effects of *Eubothrium salvelini* (Schrank, 1790) on sockeye salmon, *Onchorhynchus nerka* (Walbaum), in Babine Lake, British Columbia. *J. Parasitol. 56:* 321-322.
SWEETING, R. A. 1974. Investigations into natural and experimental infections of freshwater fish by the common eye fluke *Diplostomum spathaceum* Rud. *Parasitology 69:* 291-300.
TEDLA, S. and C. H. FERNANDO. 1970. On the biology of *Ergasilus confusus* Bere 1931 (Copepoda) infesting yellow perch *Perca fluviatilis* in the Bay of Qunite, Lake Ontario, Canada. *Crustaceana 19:* 1-14.
THOMAS, J. D. 1964. Studies on populations of helminth parasites in brown trout *(Salmo trutta* L.). *J. Anim. Ecol. 33:* 83-95.
VARLEY, G. C., G. R. GRADWELL, and M. P. MASSELL. 1974. Insect Population Ecology: an Ecological Approach. Blackwell Sci. Pub., Oxford.
VIK, R. 1954. Investigations on the pseudophyllidean cestodes of fish, birds and mammals in the Ånøya Water System in Trøndelag. Part I. *Cyathocephalus truncatus* and *Schistocephalus solidus*. *Nytt. Mag. Zool. 2:* 5-51.
VIK, R. 1958. Studies on the helminth fauna of Norway II. Distribution and life cycle of *Cyathocephalus truncatus* (Pallas, 1781) (Cestoda). *Nytt. Mag. Zool. 6:* 97-110.
WALKEY, M. and R. H. MEAKINS. 1970. An attempt to balance the energy budget of a host-parasite system. *J. Fish Biol. 2:* 361-372.
WILLIAMS, H. H. 1967. Helminth diseases of fish. *Helm. Abs. 36:* 261-295.
WILLIAMS, H. H. and O. HALVORSEN. 1971. The incidence and degree of infection of *Gadus morhua* L. 1758 with *Abothrium gadi* Beneden 1871 (Cestoda: Pseudophyllidea). *Nor. J. Zool. 19:* 193-199.
WILSON, R. S. 1971. The decline of a roach *Rutilus rutilus* (L.) population in Chew Valley Lake. *J. Fish Biol. 3:* 129-137.
WOOTTEN, R. 1973. Occurrence of *Bunodera luciopercae* (Digenea: Allocreadiidae) in fish from Hanning-

field Reservoir, Essex. *J. Helminthol. 47:* 399-408.

WOOTTEN, R. 1974. Studies on the life history and development of *Proteocephalus percae* (Muller) (Cestoda: Proteocephalidea). *J. Helminthol. 48:* 269-281.

WUNDER, W. 1929. Die *Dactylogyrus* Krankheit der Karpenbrut, ihr Ursache u. ihre Bekamfung. *Ztschr. Fischerei. 27:* 511-545.

The Role of Arrested Development in the Regulation of Nematode Populations

GERHARD A. SCHAD

Department of Pathobiology
University of Pennsylvania

INTRODUCTION

Contrary to the widely prevailing view that parasitic nematodes in their normal definitive hosts develop to adulthood within a relatively restricted and characteristic prepatent period, it is becoming increasingly apparent that this concept is too simple, and that prolonged interruption of development is a common alternative. Many variables influence parasite development, and even in highly susceptible individual hosts, growth may be interrupted. A state of arrested development, variously called inhibition, hypobiosis or diapause, may intervene, so that, at some stage short of sexual maturity, development ceases temporarily. This is not simply the brief hiatus in growth within the normal time frame which occurs at moulting and is sometimes referred to as a lethargus, although often both processes occur at the same point on the growth curve of a species.

Arrested development of the kind under consideration here has been defined by Michel (1974). He characterizes it as a "temporary cessation of development of nematodes at a precise point in early parasitic development, where such an interruption contains a facultative element, occurring only in certain hosts, certain circumstances, or at certain times of year, and often affecting only a proportion of the worms". He also notes that when only a proportion are affected, the remainder usually develop normally, resulting in a bimodal size distribution which is characteristic of the phenomenon and sets it apart from mere retardation of growth and stunting, both of which are symptomatic of growth in hosts having an innate or acquired resistance to the parasite. At present, definitions can only be provisional, since, in this active area of research, they are likely to be subject to rapid obsolescence. For example, very recent conclusions of Waller and Thomas (1975) may require modification of the definition. These authors claim that in northeastern England *Haemonchus contortus* has but one generation per year, and that arrest at the fourth stage is a normal, obligatory part of the life cycle. If this proves to be true, Michel's definition will require modification to include the concept that when facultative arrest is replaced by obligatory arrest in local geographic strains, the replacement will occur in ecologically marginal areas within the larger range of the parasite, where external environmental conditions barely allow the continued existence of the parasite.

SIGNIFICANCE OF ARREST: CHANGING CONCEPTS

Arrest of the kind under discussion involves a *temporary* cessation of development, a fact which was not always recognized. Early in the development of experimental parasitology, investigators realized that, even in highly susceptible hosts, some fraction of any dose of worms fails to mature. A careful search in these instances often revealed the presence of larval worms long after the majority had completed development. Such larvae were considered abnormally lacking in vigor and without capacity for further development. However, in 1928, Scott showed that, although some third stage larvae of *Ancylostoma caninum* recovered from the intestine of dogs weeks after oral infection were no more developed than when they were initially given, they retained the capacity to develop further. This was demonstrated by feeding the inhibited larvae to other uninfected dogs where again part of the dose developed, but part did not, the latter surviving, instead, as inhibited larvae.

Subsequently a number of investigators, but especially veterinary parasitologists, came to consider developmental arrest a mechanism by which the densities of populations of adult worms were regulated. This was particularly the case with regard to the gastrointestinal nematodes of livestock. It was generally agreed that developmental arrest was a manifestation of acquired resistance following previous exposure to parasitism. Later still, this immunological viewpoint was expanded to include host-parasite associations in which previously naive hosts could also regulate their adult worm burdens immunologically; in these cases it was thought that the presence of the parasite was not recognized until some critical threshold was exceeded, whereupon the host reacted to control the parasite's biomass. Developmental arrest was considered one of several forms of control.

In the recent literature, there is a growing trend to view the function of arrested development in some species, or in particular geographic variants of these species, as primarily a mechanism permitting parasites to survive during seasons unfavorable for their external development as free-living larvae. From this standpoint, developmental arrest is a secondary life history option ensuring survival from year to year and/or transmission to new groups of young susceptible hosts. It is a normally occurring alternative life history pattern which has evolved when species or strains are confronted seasonally with long periods of harsh environmental conditions.

While it is certainly true that, in these cases, seasonally occurring developmental arrest permits survival during periods of adversity, it is equally apparent that the entry of newly acquired parasitic nematodes into arrest, often at precisely the time when adult worm burdens are becoming particularly dense, is a highly adaptive mechanism for regulating populations of adult worms. Furthermore, when the external environment is alternately favorable and unfavorable for free-living development, it follows that the availability of the infective stage usually will maximize at the end of the favorable season. Thus, the host will be exposed to the most intense invasion at this time, when often for the host, too, the environment is beginning to degenerate. Hence, larval arrest would, in these instances, limit the abundance of adults relative to the less pathogenic larvae during the most stressful part of the year. It appears, therefore, that the trend to interpret developmental arrest primarily as an alternative life history pathway circumventing adverse external conditions is excessively restrictive. A high density of adult worms is an unfavorable component of a parasite's environment. When high density is considered a form of adversity, then a propensity to arrest at the times when transmission maximizes is equally adaptive, whether viewed as a mechanism to limit the size of the adult worm population or as a mechanism to permit survival during seasonally unfavorable, external, environmental conditions.

FACTORS WHICH INDUCE ARRESTED DEVELOPMENT

The factors known to induce arrested development among nematodes may be grouped into three major categories: (1) external environmental factors which induce a potential for a diapause-like state subsequently expressed within the host, (2) host factors which determine the host's suitability as an environment for further development, and, when adverse, lead to arrest, and (3) parasite-related factors, either genetic or density-dependent, which, either alone or together with factors already listed, induce arrest. To emphasize the concept that factors from the three groups probably can and do act together to induce arrest, a Venn diagram, linking these factors, is presented (Figure 1).

ENVIRONMENTALLY INDUCED ARREST

The population structure of the gastrointestinal nematodes occurring within grazing animals in temperate

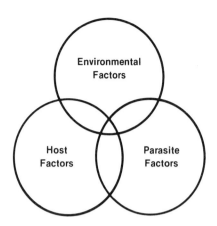

Figure 1. *Major groups of factors known to induce a potential for developmental arrest in nematodes parasitic in animals.*

parts of the world shows a marked seasonal variation. During spring and summer, when free-living stages can develop, hosts harbor egg-laying adult worms, whereas, in winter, when conditions for external larval development are unfavorable, their nematode populations consist largely of arrested larvae. Overwintering in the host has been the subject of recent careful epidemiological investigations in Canada (Ayalew and Gibbs, 1973), England (Connan, 1971), Scotland (Reid and Armour, 1972), Poland (Malczewski, 1970) and elsewhere. Table I lists some species which arrest within their hosts during seasons unfavorable for free-living development. In each case, there is good evidence for a marked seasonal change in population structure based on either periodic collections of parasites at necropsy or on comprehensive longitudinal epidemiological studies. Figure 2, based on data presented by Ayalew and Gibbs (1973), shows the kind of seasonal variation in population structure found by seasonal examination of animals at necropsy.

In temperate areas where winters are mild, but summers are hot and dry, arrest within the host may occur during summer (Hotson, 1967; Anderson, 1972; Shimsony, cited in Armour and Bruce, 1974). It is probable that seasonal arrest also occurs in response to unfavorable external conditions in the tropical parts of the world (Schad, et al., 1973; Bindernagel and Todd, 1972), but this phenomenon has received comparatively

TABLE I.

Some nematodes known to arrest within their definitive hosts during seasons unfavorable for their free-living development.

Species	Host	Area	Season	Stage	Source (see literature cited)
Ancylostoma duodenale	Man	India	Dry	L3	(75, 87)
Chabertia ovina	Sheep	England	Winter	L4	(21)
Cooperia oncophora	Cattle	Canada	Winter	L4	(90)
Haemonchus contortus	Sheep	Canada, N. Eur., S. Afr., N.Z. & Aus.	Winter	L4	(8, 37) (20, 22, 54) (73) (17, 55, 42)
Hyostrongylus rubidus	Swine	England	Winter	L4	(19)
Nematodirus spp.	Sheep	Canada, Scotland	Winter	L4	(8) (76)
Ostertagia ostertagi	Cattle	Canada, N. Eur., Australia, Israel	Winter, Hot, Dry	L4	(90) (54, 82) (1, 47) (4)
Ostertagia spp.	Sheep	Canada, Scotland	Winter	L4	(8) (76)
Oesophagostomum columbianum	Sheep	Australia	Winter	L4	(62)
Dictyocaulus viviparus	Cattle	Europe, Canada	Winter	L5	(99) (44)

little attention.

Traditionally, arrested development has been considered one of the manifestations of immunity to nematodes (Donald, et al., 1964; Kelly, 1973; Soulsby, 1966), and most attempts to explain the cause of seasonal developmental inhibition were based on changes in host resistance which occurred during the year. However, recently it was shown that arrested worms can occur seasonally in young, highly susceptible hosts with no previous exposure to infection; thus, while under some circumstances, immunity can play an important role in the induction of arrest, it does not play a primary role in the seasonally occurring arrest under discussion in this section of this paper.

External environmental factors were first proposed as stimuli for arrest by Anderson, et al. (1965) with regard to the stomach worm of cattle, *Ostertagia*

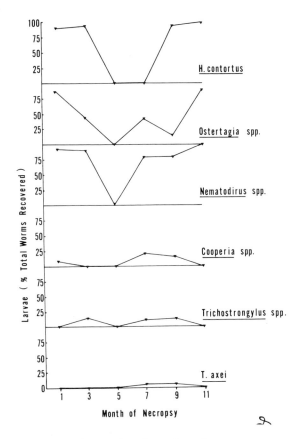

Figure 2. Seasonal variation in the structure of the nematode populations in sheep in eastern Canada. Arrested development occurs in Haemonchus contortus, Ostertagia *spp. and* Nematodirus *spp. but not in the remaining genera (from data presented by Ayalew and Gibbs, 1973).*

ostertagi. Using what has since become a familiar experimental technique involving so-called "permanent calves" and "tracer calves", they demonstrated that immunity could not be the primary inducing factor, but rather that seasonal factors, acting either directly on free-living stages of the parasite or indirectly via the host's hormones, induced arrest. The "permanent calves" grazed and maintained the natural contamination of a pasture with the free-living stages of *O.*

ostertagi. Periodically, during spring and summer, groups of young, worm-free "tracer calves" were introduced to the contaminated pasture. Each group of tracers grazed for 14 days, it being considered unlikely that, in this short period, an acquired immunity to *O. ostertagi* could develop. After the period of grazing, each group of calves was housed so as to preclude the acquisition of additional worms; those worms which had been acquired on pasture were given time to develop to adulthood. In this way, it was shown that during spring and summer the worms developed to adulthood promptly, whereas in autumn the development of a large proportion was interrupted early in the fourth larval stage. In a continuing series of investigations, Armour and his colleagues have confirmed and extended these studies. By grazing tracer calves for one or for 14-day periods from spring through autumn, Armour, Jennings, and Urquhart (1969a) demonstrated that the degree of arrest (approximately 70% in autumn) was independent of the size of the worm burden and of the length of time over which calves were permitted to graze. Some calves, restricted to but one day of grazing in November, harbored only 1400-1500 worms at necropsy and these worm populations were constituted mostly of arrested larvae (80-86% of the population). This investigation contributed further evidence that the immune status of the host did not play an important role in the induction of arrest in this species.

It was also shown that a seasonal environmental factor(s), acting directly on the free-living stages of the parasite rather than on the host, and expressed through its endocrine system, was the stimulus for developmental inhibition (Armour, Jennings, and Urquhart, 1969b). Comparisons involving an old laboratory strain (the Weybridge strain) and a freshly isolated wild strain demonstrated that the capacity to be induced to arrest could be lost in the laboratory, presumably through rapid passage which selected against the arresting trait.

Subsequent investigations showed that the primary stimulus for arrest was exposure of infective larvae to cold, and that this seasonal form of inhibited development was a diapause-like phenomenon (Armour, 1970; Armour and Bruce, 1974; Bruce and Armour, 1974).

TABLE II.

Experimental models for investigations of environmentally stimulated developmental arrest among nematodes parasitic in animals.

Species	Origin	Host	Stimuli Investigated	Result	Source
Obeliscoides cuniculi	Canada	Rabbits	Low Temp., Δ Temp.	+	(36, 48)
Ostertagia ostertagi	Scotland/ England	Calves	Low Temp., Δ Temp.	+	(4, 66) (66)
Cooperia oncophora	England	Calves	Low Temp., Δ Temp.	+ +	(66) (66)
Haemonchus contortus	Canada	Lambs	Δ Temp. & Photoperiod	+	(14)
			Short Photoperiod	+	(37)
	England		Low Temp., Δ Temp. & Photoperiod	−	(22)
	New Zealand		Low Temp.	−	(55)
Dictyocaulus viviparus	Canada	Calves	Low Temp.	+	(44)
Ancylostoma caninum	Maryland	Pups	Short photoperiod, Low Temp., Δ Temp.	− ± +	(86)

Less than 1% of either freshly cultured infective larvae or larvae stored for 24 days at 18-20°C became dormant when dosed to calves, whereas 44% of larvae stored at 4°C were inhibited. Storage of larvae at 4°C for 8 weeks induced 66% arrest, but after long storage (19-46 weeks), progressively fewer larvae arrested. Armour and Bruce (1974) attribute this reversal to the death of inhibition-prone larvae after prolonged storage.

Michel and coworkers (1973; 1974; 1975a) have confirmed that the primary stimulus for arrest can be an environmental signal. Infective larvae stored at constant low temperature (4°C) were more rapidly conditioned to arrest than larvae stored at a higher constant temperature (15°C), but a sudden decline in temperature (15 ↓ 4°C) increased the percentage of larvae which would arrest very rapidly (Figure 3).

Michel, Lancaster, and Hong (1973) could not confirm that the loss of the capability to arrest occurring after prolonged storage was attributable to differential mortality, i.e. due to the death of those larvae which responded to low temperature stimulation. In

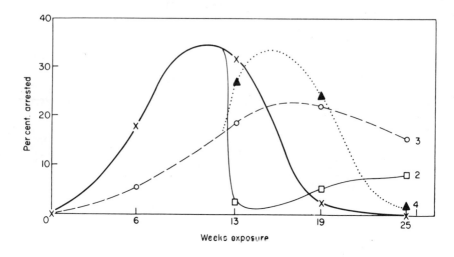

Figure 3. Mean percentage arrest in 4 groups of Ostertagia ostertagi *larvae exposed to different temperatures or changes in temperature during storage and then dosed to calves. Groups 1 and 3 stored for 25 weeks at $4°C$ or $15°C$, respectively. Group 2 stored at $4°C$ for 12 weeks and then at $15°C$, and Group 4 stored at $15°C$ for 12 weeeks and then at $4°C$. Reproduced from Michel, et al. (1975).*

their investigations, the infective larvae lost their potential to arrest spontaneously beginning after 3 months of storage at $4°C$, and, after 5-6 months, virtually none arrested; this loss could be accelerated sharply by exposing the cold-conditioned larvae to a sudden increase in temperature. When larvae which had been stored at $4°C$ for 12 weeks were held at $15°C$ for one week, only 2% arrested.

Michel, Lancaster, and Hong (1973) have also demonstrated that selection for or against the ability to arrest can be very rapid, so that strains may vary from farm to farm, depending on management practices. Presumably this could generate a mosaic pattern of sympatric forms, each with its own developmental characteristics within a small geographic area. Among parasites of livestock, such local variation could obscure the major geographic clines in the ability to arrest which one would expect to exist, and which would adapt the parasite to geographically varying external environmental conditions.

Using the same calves in which they investigated the induction of arrest in *Ostertagia ostertagi*, Michel, Lancaster, and Hong (1975a) examined the phenomenon in *Cooperia oncophora*, an intestinal trichostrongyle of cattle. There was remarkably little difference between the response of the two species.

Obeliscoides cuniculi, a parasite of rabbits and a close relative of *Ostertagia* spp., also overwinters in its host (Gibbs, personal communication). Like *O. ostertagia*, the ability to interrupt its early parasitic development is induced by maintenance at low temperature (Fernando, Stockdale, and Ashton, 1971; Hutchinson, Lee, and Fernando, 1972). This process is accelerated by a sudden decrease in temperature ($15 \downarrow 5°C$) and again is reversible.

The free-living stages of *Haemonchus contortus* fail to survive the winter in cold climates, and the life span of the adult worms within the host is short; thus, for this parasite to survive in marginal areas of its range, it has had to evolve a mechanism for overwintering. Remaining arrested in the host during the winter is such a mechanism. Several authors have reported that a full 100% of the winter worm burden occurs as dormant, fourth stage larvae (Ayalew, et al., 1973; Ayalew and Gibbs, 1973; Waller and Thomas, 1975).

Winter dormancy in this species has been reported in numerous north and south temperate areas, including the United States and Canada (Benz and Todd, 1969; Blitz and Gibbs, 1972a; 1972b), England (Connan, 1968), Poland (Malczewski, 1970), South Africa (Muller, 1968), Australia (Gordon, 1973) and New Zealand (Brunsdon, 1973). In the tropics, where dryness rather than cold acts as a seasonal barrier to transmission, parasitic development is interrupted during the dry season (Hart, 1964), but this phenomenon has received little attention.

The initial suggestion that seasonal developmental arrest in *H. contortus* might be attributable to the effects of changing external environmental conditions acting on the free-living larvae came from the work of Connan, who showed that, in England (Cambridge), larvae arrested in parasite-naive, tracer lambs in late summer and fall. By October, all newly ingested larvae became arrested. Blitz and Gibbs (1972a), working near Montreal, Canada, found that when lambs grazed naturally contaminated pasture in September, 100% of the *H. contortus* larvae ingested became arrested. In lambs dosed with larvae cultured to infectivity in the laboratory but held thereafter under September pasture conditions, 96% arrested. In contrast, larvae taken from the same original culture and held in the laboratory under constant environmental conditions (21°C, constant light), showed little tendency to arrest (10% and 40% in each of 2 sheep). At that time of year, the hours of daylight decreased from about 14 to 12.5, and the mean temperature from 17° to 11°C. Blitz and Gibbs, like Armour and his colleagues, concluded that seasonal arrest in trichostrongyle nematodes was a diapause-like phenomenon. Subsequently, Gibbs (1973) compared the arrest inducing ability of a short (10 hr) or a long (16 hr) photoperiod acting on larvae stored at 20°C for 6 weeks. Arrest among larvae which had been exposed to the short photoperiod exceeded that occurring among larvae used fresh from culture, or used after exposure to long photoperiods, by 70%. The details of these studies have not been published, but they suggest that the Canadian strain is photosensitive.

British strains apparently respond differently. Connan (1975) attempted to repeat the investigations of Blitz and Gibbs under conditions prevailing near Cambridge. Infective larvae cultured in the dark at 25°C were harvested and stored on damp filter papers in Petri dishes contained in sealed polyethylene bags. These were then subjected to pasture conditions for increasing 10-day intervals up to 40 days. Other larvae were stored for the 40-day period either inside, where temperature (4°C) and photoperiod (L:D 0:24) were constant, or outside, but in constant darkness, so that they experienced temperature changes only. Finally, one batch of larvae not subjected to any of the storage (aging) conditions outlined above was given to lambs when fresh from a two-week old culture. Three worm-free lambs were infected with larvae of each experimental treatment. Little difference was observed in the degree of developmental inhibition between larvae of the different treatment groups. Indeed, 91-97% of the fresh larvae became inhibited, leading Connan to suggest that unidentified environmental factors acting during the 12 days in culture were capable of inducing a potential for arrest. That a short photoperiod might be the factor which induces arrest in this strain as in the Canadian strain does not appear to have been excluded. Although the necessary details are not presented without ambiguity, all groups of larvae were apparently cultured in the dark and then aged either under November-December photoperiods or in the dark. Apparently in other experiments designed to test the effects of low temperature only, larvae were not reared in the dark, and smaller percentages arrested. It would seem, therefore, that short day length cannot be ruled out, and may even have been the important factor.

Waller and Thomas (1975), working in northeast England, explicitly discount both photoperiod and temperature as the major stimuli for arrest. They point out that, since arrest exceeding 50% was observed in July, chilling could not be a stimulus. Presumably they exclude photoperiod from consideration, since the larvae observed to be arrested at this time of year would have been on pasture when day length was at its maximu. Although Gibbs (1973) did find that a Canadian strain is stimulated to arrest by storage under short day conditions, this does not mean that all geographic races of *Haemonchus* use an autumnal day length as a signal to arrest. The signal to enter diapause may be received well in advance of the time when an organism enters this state; this has often made it difficult to associate such signals with the triggering of the phenomenon (Wigglesworth, 1970).

HOST FACTORS

Among nematodes, developmental arrest attributable to either natural or acquired host resistance is well known. Attributes of hosts which contribute to their natural ability to resist nematode infections and, hence, to the induction of arrest, are age, sex and species. In these instances, however, it is important to distinguish clearly between mere stunting and true arrested development. This, as Michel (1974) emphasizes, has not always been done; indeed, under some circumstances, it is difficult to make the distinction. For the purposes of this review, stunting is considered to be characterized by a retarded rate of growth which produces an adult worm of subnormal size and fecundity. This phenomenon is not of direct interest here. Arrested worms, on the other hand, have grown at a normal rate to a particular species-specific point at which arrest characteristically intervenes. The inhibited worms show an appropriate degree of development for their characteristic arresting stage.

While the great majority of investigations dealing with seasonally induced arrest have been concerned with parasites of ungulates of economic importance, investigations dealing with age, and particularly with sex and arrest, have involved mainly laboratory and companion animals. For obvious economic reasons, experimentation involving large farm animals has usually been done with either lambs, ewes or calves, especially bull calves. Thus, while age and sex of the host probably do influence arrest of nematodes in ungulates, these factors have not been considered of particular importance in these host-parasite associations. In contrast, developmental arrest due to resistance conferred by age or sex of the host is well known among the nematodes of dogs. In young pups, infections with the common dog hookworm, *Ancylostoma caninum*, mature and become patent within about two weeks. Older animals, even though they have not been infected previously, become resistant, and fewer larvae mature. Incoming infective larvae follow the somatic migratory route and become dormant in the muscles. Thus, Miller (1965) observed that resistance, as judged by the failure of larvae to develop into adult worms within 21-26 days, was manifested in helminth-naive 8-month-old dogs, and that in females it exceeded that observed in males of comparable age. Taking infections established in three-month old pups as a baseline, he showed that infection levels were reduced by 10% and 51% in 8-month old males and females respectively, and by 53% and 81% in corresponding groups of 11-month old animals. Until recently, the storage of larvae in the musculature of older dogs, particularly bitches, was an inference based on the observation that transmammary transmission of *Ancylostoma caninum* occurred regularly, even when the dam had been protected from infection prior to and during lactation. That the specific site of storage was muscle tissue was an assumption based on larval behavior in laboratory animals. Recent investigations have confirmed these concepts. Lee, Little, and Beaver (1975) have demonstrated *A. caninum* larvae in the muscles of five and nine infected dogs, and in an elegant series of experimental studies, Stoye (1973) confirmed that the inferred sequence events actually occurs. This author infected a series of helminth-naive bitches with *Ancylostoma caninum* and subsequently demonstrated third stage larvae in their musculature. These survived in the muscles for at least 240 days. Larvae could be found in the mammary glands only during lactation, and appeared in samples of milk in greatest abundance during the first week after parturition. Bitches infected on one occasion only (20,000 larvae) excreted larvae in their milk after each of three consecutive pregnancies, the output of larvae declining from one post-parturient period to the next. Larval excretion was experimentally induced by treating bitches with oestradiol or oestradiol and progesterone.

Similar events occur in canine toxocariasis (Webster, 1956; Greve, 1971; Sprent, 1958). In young pups, larvae undergo tracheal migration and mature in the intestine. In dogs exceeding six weeks of age, progressively more worms follow the somatic route and encyst in the tissues. Thus, Greve, using helminth-naive beagles, demonstrated that few arrested larvae occurred in the tissues of pups experimentally infected at three weeks of age, and that the larvae migrated in synchrony to the intestine. In 3-month and in year-old dogs, few larvae reached the intestine and developed successfully; rather they occurred in ascaridial granulomata in the tissues. There were insufficient year-old dogs infected to examine the data for sex-related differences, but at three months of age, the sexes did not differ with regard to the presence of intestinal adult worms or the abundance of larvae in the tissues. However, it is well known that the sexes differ with regard to the presence of adult worms. For instance, Ehrenford (1956) observed that intestinal, adult-worm infections were three times as prevalent in males as in females and confirmed that the sex-related difference in prevalence increased with age.

Interest in visceral larval migrans has prompted considerable investigation of the development of these species in abnormal hosts. Indeed, it is clear that paratenic hosts (small mammals) are probably a source of infection for free-ranging dogs. Miller (1970; 1971) has shown that hookworm larvae survive in the musculature of mice for the life of the mouse, and Stoye indicates that these larvae from paratenic hosts and those from the muscles of the bitch are morphologically identical. Furthermore, he notes they are characterized by a massive accumulation of fine granular material in the intestine such as is found in physiologically aged larvae. Rogers (1939) considered this to be an excretory product. In view of Blitz and Gibbs' (1971) observation that characteristic crystalloids accumulate in the gut of arrested *Haemonchus*, one wonders if a related phenomenon may not be occurring in this case. This might be an indication that arrest has a similar physiological basis in an abnormal host species and in a normal definitive host species when it is the result of age and/or sex-related resistance.

Since, in hosts which have been exposed to infection repeatedly, a large proportion of the nematode population frequently is found arrested, many authors have considered inhibited development a manifestation of acquired immunity (Donald, et al., 1964; Dineen, et al., 1965; Kelly, 1973; Michel, 1968; Soulsby, 1958). Recently, however, Michel (1974) has argued that acquired immunity is not one of the primary causes of arrested development as was once thought, and that data traditionally interpreted as supporting an immune basis for arrest lend themselves to other equally probable interpretations. He states that "in many cases, conclusions regarding the cause of arrested development prove on close examination to be unfounded because they are based on the proportion of the worms, present in the animals at the end of an experiment, which are arrested. Since, characteristically, arrested worms persist in the host for longer than developing or adult worms, factors and circumstances which lead to a rapid loss of adults are mistakenly identified as a cause of arrested development." Thus, since host resistance is also manifested in the loss of adult worms, the mere fact that resistant hosts harbor disproportionately large burdens of arrested larvae does not prove that this factor induced arrest. While Michel is correct in indicating that data of the kind described do not constitute proof, it will be shown later that loss of arrested worms may also occur, so that, in some situations, the net effect may be attributable to immune induction of arrest counteracted by an increased rate of larval loss. Furthermore, there are so many valid examples of immunologically induced arrest that it seems premature to reinterpret the other experiments which can be reevaluated along Michel's line of reasoning.

A particularly well-known investigation in which developmental arrest was attributed to the immune response of the host involved *Nematodirus spathiger* infections in lambs (Donald, et al., 1964). Donald, Dineen and colleagues reasoned that in a highly evolved, well-adjusted host parasite relationship, the host would fail to respond immunologically to the presence of the parasite at densities of the latter which did not exceed some critical, nonpathogenic threshold. At higher densities (biomasses), it was expected that the host would respond to limit the worm population, and that previous experience with the parasite would lower the threshold at which a response occurred. Based on these concepts, a group of 30 lambs was divided into three groups of 10. Each lamb of one group (Group I) was given 50,000 infective larvae in a single dose, this number representing what appeared to be the maximum tolerated worm burden from previous reports in the literature. Lambs of Group II were given 130,000 larvae at a single exposure, a dose thought to greatly exceed the hypothetical threshold. Finally, lambs of a third group again received 50,000 larvae each, but in consecutive daily doses of 2,000 larvae for 25 days. Necropsies conducted at 40 days and 74 days after infection (or first infection in Group III) demonstrated that in terms of absolute numbers, least arrested larvae occurred in Group I, and the most in Group III, whereas Group II occupied an intermediate position. Stated as a percentage of worm burden, fourth stage larvae for Groups I, II and III respectively averaged: 20, 37 and 75% on day 40 and 20, 85 and 95% on day 74. The authors concluded that, while the host regulated its helminth populations in several ways (rejection of infective larvae, rejection of adult worms (particularly females), and interference with egg production), retardation of development in the fourth larval stage was the major factor. This point is made especially clear in Dineen, et al. (1965).

In additional experiments of similar design, Dineen, Donald and colleagues came to similar conclusions. Thus, Dineen, et al. (1965), in an experiment with *Haemonchus contortus*, divided 57 helminth free lambs into three groups of 19 and dosed them with 3000 infective larvae, 9000 larvae or 3000 larvae given in daily doses of 100. In the paper cited, only the first and third group are discussed, since almost all lambs in the second group died of haemonchosis. Again, as with *N. spathiger*, the proportion of arrested larvae and the absolute number of arrested worms was greater in the group infected daily than in the group given the same total number of larvae at one time. The fact that the absolute numbers were different obviates the possible objection that differential rates of loss of adult worms could account for the results obtained when percentages were used for between group comparisons of arrested development. The worm counts are shown in Figure 4.

Herlich (1967), in investigations of *Cooperia pectinata* which compared the parasite's development in calves given either a single infection with 30,000 larvae or the same number of larvae in 10 consecutive daily doses, also found that animals given the divided dose harbored the greater number of arrested larvae, whether judged as a proportion of worm burden or by actual worm count. Hence, it is difficult to interpret the results in any way other than that host resistance, increased by repeated sensitizing exposures, contributes to arrest.

An experiment involving another approach, namely immunosuppression, is of particular interest. Dunsmore (1961) infected each sheep in four groups of three with 100,000 larvae of *Ostertagia* spp. Beginning five days before infection, one sheep of each trio was treated

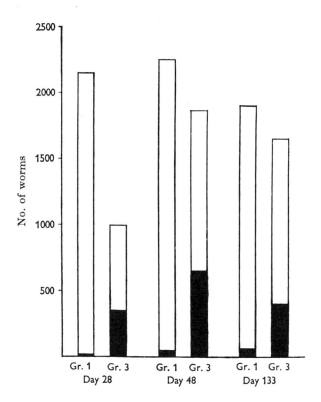

Figure 4. Structure of the Haemonchus contortus population in 2 groups of experimentally infected lambs. Lambs of Group 1 (Gr. 1) received 3,000 infective larvae each as a single dose; lambs of Group 3 (Gr. 3) also received 3,000 larvae but in daily doses of 100. Open bars show the mean number of adult worms; solid bars show the mean number of larvae recovered. Reproduced from Dineen, et al. (1965).

with cortisone, and another subjected to whole body irradiation, while the third served as an untreated control. Fourteen days after infection the sheep were slaughtered and differential worm counts were made. The control sheep harbored many more arrested worms in absolute numbers than did those which had been immunosuppressed; percentage arrest was 8.5 in the cortisone-treated sheep, 18.7 in the irradiated group, and 43.2 in the controls.

PARASITE-RELATED FACTORS

Genetic

In the discussion of environmentally induced arrest, presented earlier in this paper, it was noted that, in some localities, *Ostertagia ostertagi* arrests in the host during cold months, whereas, in other areas having mild winters but summers that are too hot and dry for the survival of free-living larval stages, arrest occurs during the latter period. For example Smeal (cited in Kelly, 1973) has observed this response in parts of Australia. Presumably *O. ostertagi* has become adapted to local conditions throughout its vast range and a myriad of strains exist.

It is also apparent that strains of *Ostertagia* may rapidly lose their ability to be induced to arrest by environmental stimuli. Investigators at Glasgow discovered that a wild strain which had given rise to unseasonable disease (type II or winter ostertagiasis) and which could, therefore, be considered arrest prone, could be induced to arrest by exposure to autumnal conditions, whereas a laboratory strain (the Weybridge strain) which had been passaged rapidly in calves, failed to arrest after exposure to the same conditions. Passage of the latter when fecal egg counts maximized early in the course of infection had apparently selected against the arresting trait.

Recent investigations by Michel, Lancaster, and Hong (1973) have been concerned with selection for arrest. They have found that, beginning with the Weybridge strain and by strong selection for arrest, the degree of change which had occurred over 12 years and 28 generations could be reversed in a single generation.

With regard to *Haemonchus contortus*, Blitz and Gibbs (1972a) implicated Canadian autumnal climatic conditions in the induction of arrest of a strain from southern Quebec. A later publication of Gibbs (1973) reported that exposure of infective larvae to a 10-hour photoperiod acted as a stimulus for diapause. Conan (1975) subsequently stated that neither short photoperiod nor low temperature induced arrest in a strain from Cambridge, England. However, as indicated earlier, it is not completely clear that photoperiod can be excluded, but if one accepts Connan's own conclusions, the basis of arrest in his English strain differs markedly from that of a Canadian strain. This is not surprising, given the plasticity of *H. contortus* (Whitlock, 1966) and the different selection pressures that are probably operative under Canadian as opposed to British conditions.

That clines exist in capacity for arrest associated with the severity of winter was suggested by Gordon (1974), who presented preliminary investigations on Australian strains of *H. contortus*. A strain from the colder, southern part of the continent had a marked capacity for arrest, a strain from the subtropical area near Brisbane lacked this ability and a strain isolated from an intermediate area gave an intermediate response.

Geographic races of insects often differ sharply in their capacity for diapause depending on the severity of the unfavorable season. At the extreme margins of its range, some races of a species may encounter such short growing seasons that only one generation per year can be completed, whereas several may be completed elsewhere. The former are said to be univoltine and, in these peripheral, univoltine races, diapause is sometimes genetically determined rather than environmentally induced. This is, in fact, what Waller and Thomas (1975) have suggested to be true for *Haemonchus contortus* in northeastern England. However, based on entomological experience, it is probably too early to come to this conclusion. Either increasing, or decreasing day length, can signal impending winter, depending on the length of the period between the organism's receipt of the stimulus for diapause and its entry into this state. Therefore, it may be premature to conclude that the strain of *Haemonchus* from northeastern England has a genetically fixed prepatent period of 9-10 months.

Nawalinski and Schad (1974) suggested that an Indian strain of *Ancylostoma duodenale* isolated in West Bengal where, for about eight months annually, it is too dry for free-living development of the species, might have a genetically fixed, long, prepatent period. This suggestion was offered among several, since, in two self-induced human infections, a full 100% of the inoculum arrested, although nothing remarkable had been done either during culture or subsequent maintenance of the larvae. These were reared in the dark (i.e. in an incubator) at 28°C and maintained under these conditions until required for infection. On the other hand, the strain had been isolated by culturing the eggs extracted from a <u>single</u> female worm, which suggested that a genetic factor might be important. However, when Schad and Nawalinski (unpublished) cultured larvae in either continuous darkness or with daily exposure to light and dosed these to pups, larvae of both groups developed promptly. Since, if anything, one would expect arrest to be more probable in an abnormal host, it is apparent that the Bengal strain of *A. duodenale* does not have a genetically determined, obligatory 40-week prepatent period, as we once, among other alternatives, suggested.

It is clear from the information just reviewed that, in species of parasitic nematodes with extensive geographic ranges, local populations exist which may or may not respond to environmental signals to arrest. Responsiveness correlates with the severity of climate and thus closely parallels the situation known to occur among the insects which show geographic variation in propensity for diapause. Although it has been suggested that genetically determined univoltism may occur in geographic races of nematodes which are adapted to particularly short growing seasons, it appears that this interesting suggestion requires stronger evidence than presently available before it can be accepted.

Adult Worms

The presence of adult worms has been thought to play a role in larval arrest ever since 1953, when Gibson showed that treatment of housed horses with phenothiazine produced only a temporary elimination of their small strongyles. In an experiment lasting for three years, 6 horses were stabled under conditions designed to prevent reinfection. After fecal egg counts were begun, the animals were treated with phenothiazine, a compound known to be very effective against the lumen dwelling adults of the small strongyles and moderately effective against similar stages of large strongyles, but ineffective against the tissue dwelling larvae of both groups. After treatment, egg counts declined sharply. Several weeks later, however, eggs again became relatively numerous in the feces, and thereafter they increased gradually until the horses showed substantial fecal egg counts. The horses were then retreated. Each time, Gibson showed that numerous worms were discharged with the feces, and egg counts fell to zero or to negligible numbers. By using egg counts, along with larval culture and differential counts of the larvae produced, he was also able to show that the first eggs to reappear were those of the large strongyles, and that it was several weeks before those of small strongyles were in evidence. Over the long term, after each treatment, the egg count tended to return to a lower level and, generally, the time interval until eggs were again abundant in the feces, increased.

Gibson concluded that, since reinfection could not have occurred given the conditions under which the horses were maintained, the observations could be explained in two ways: 1) either only a fraction of the adult worm burden was removed after each treatment and oviposition was temporarily inhibited among the remaining worms, or 2) the adult worms had been removed, but, each time they were removed, a new population of larvae was reactivated and recruited from the reservoir of inhibited larvae occurring in the mucosa.

Gibson rejected the first of these explanations, since it could only explain some of his observations. Partial clearing of the large strongyles along with a post-treatment reduction in their egg output was not only possible, but was the probable explanation for the first eggs to reappear in the feces. Indeed, fecal cultures and differential counts of the resulting larvae had shown that the first eggs to reappear were those of the large strongyles. However, after each treatment, large numbers of small strongyles were expelled, and he knew from a previous investigation (Gibson, 1950) that phenothiazine, at the dosage given, removed 100% of the lumen-dwelling adults, without affecting the histotropic larvae. Thus he proposed that it was, in fact, the adult worms that caused the larvae to remain inhibited in the mucous membrane of the caecum. This was a bold hypothesis for its time, since it was still widely believed that histotropic larvae of nematodes which did not return to the lumen promptly and continue their maturation, were destined to die without further development. This paper, then, made a considerable impact, and it continues to be widely accepted that removal of adult worms can provoke resumption of development among dormant larvae.

It has not been clear, however, whether the prior presence of adult worms causes larvae to become arrested, or whether, once larvae are arrested, the adults tend to keep them in arrest. In any case, the generality of the concept has recently been called into question by Michel (1974), who, on the basis of his work with *Ostertagia ostertagi*, believes that gastrointestinal nematode populations of grazing animals are constantly turning over, adults being lost and replaced by worms from the reservoir of arrested larvae. Michel suggests that the larvae which replaced the adults in Gibson's horses would have done so in any case. Close inspection of Gibson's data shows that, after treatment, it could take five months or more before the eggs of small strongyles reappeared in the feces in abundance, and after several treatments, the interval until the eggs of these worms appeared in abundance became increasingly long. These observations, then, seem to provide strong support for Michel's view. If the adult worms were, in fact, holding the larvae in a suppressed state, many worms should have resumed development synchronously so as to cause a prompt, synchronous increase in egg count.

Gibson's paper, as was indicated, generated considerable interest, and in Australia it was suggested that in *Haemonchus placei* infections, adult worms suppress the development of larvae (Roberts and Keith, 1959). It was further suggested that it might even be possible to inadvertently kill an infected animal by worming and precipitating a release of larvae in sufficient numbers to overwhelm the host. Michel (1952a; 1952b), working with *T. retortaeformis* in rabbits as a laboratory model system for trichostrongyle infections in ruminants, had reported the natural occurrence of this phenomenon. He found that *T. retortaeformis* adults are expelled from the rabbit in waves, and that following each such expulsion, a new batch of larvae is released from dormancy. Occasionally a new batch of sufficient abundance to be lethal was released.

Dunsmore (1963) attempted to generate quantitative data supporting the idea that adult worms keep arrested larvae suppressed and, to this end, conducted two experiments with *Ostertagia* spp. He infected sheep, treated them with phenothiazine to selectively remove the adults, and then examined them for evidence that resumption of development had been triggered by this removal. In the first of his investigations, the anthelimintic was less effective than one would have expected, and results were inconclusive. In the second experiment, no difficulty was experienced and good evidence was obtained in support of the concept that adult worms hold larvae dormant. Each of 10 worm-free lambs had been given 40,000 *Ostertagia* larvae and, after 25 days, was treated with phenothiazine. Two days after treatment, half of the lambs were sacrificed and seven days later those remaining were slaughtered. In this way Dunsmore proposed to determine the parasite's population structure immediately after treatment had had its effect, and to document any changes caused by the loss of adult worms. The two parasite populations, two days and seven days after treatment respectively, were constituted as follows:

Arrested 62.6%, developing 0.4%, adult 37.0%.
Arrested 3.3%, developing 10.7%, adult 86.0%.

Dunsmore concluded that removal of the adult worms triggered development among those arrested, so that their population declined sharply, and that this was reflected in increased percentages of developing and adult worms.

In a thought provoking paper in which Michel first strongly advocated the concept that the gastrointestinal nematode populations of ruminants are being constantly turned over, he also suggested that adult worms play a part in the induction of larval arrest. Michel (1963) infected several groups of calves daily with 1500 larvae of *Ostertagia ostertagi*, but one group was treated once weekly with an anthelmintic (Neguvon) known to be very effective against adult *O. ostertagi* but completely ineffective against all immature stages

of the parasite (Banks and Michel, 1960). Both adult
worms and arrested larvae failed to accumulate while
the anthelmintic was being given, whereas arrested
larvae did accumulate in a comparable group of calves
dosed at the same rate with larvae from the same batch,
but not treated to remove adult worms. Furthermore,
when, at 168 days after the initiation of the experiment, the anthelmintic treatment was discontinued, a
small population of adult worms built up along with a
large population of early fourth stage larvae. Michel
concluded that, as long as the adults were being removed, incoming larvae were not inhibited in their development, and after reaching adulthood they themselves were
removed by the anthelmintic. When, on the other hand,
the adults were allowed to remain, incoming worms were
inhibited in their further development and shunted into
the reservoir of arrested larvae (Figure 5). Michel

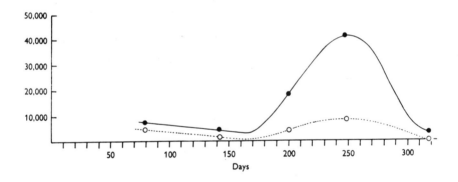

Figure 5. Growth curve of Cooperia curticei *in sheep.
Open circles = fourth stage females, solid circles
fourth stage males; + = adult males; x = adult females;
triangles = sexes not differentiated. Reproduced from
Sommerville (1960).*

believes that adult *O. ostertagi* are "in some way
necessary to inhibition of development", but that only
few are necessary and that no quantitative relationship
exists between the number of adults and the extent to
which arrest is induced among incoming worms. Indeed,
in the three experimental groups constituting this
experiment, maximal rates of inhibition were observed

while adult populations were small.

Subsequently, Michel (1971) specifically set out to test the role of adult worms in the inhibition of *Ostertagia ostertagi* in calves. Beginning with 22 calves 100 days of age, he infected each daily with 150 third stage larvae until the periodic slaughter of individual calves suggested that populations of arrested worms could be increased no further and that they had, in fact, begun to decline. The remaining 15 calves were then divided into five groups of three, and on the twentieth day after daily infection was begun, the experiment *per se* was initiated. Calves of one group were sacrificed to establish a baseline. Infection of calves was discontinued in two groups, while in the remaining two it was continued on the same daily schedule for another month. One of the groups which continued to receive larvae and one which did not were treated periodically with the anthelmintic, Neguvon, to remove adult worms, but leave larvae. The results indicated that, although the populations of arrested larvae peaked at about the time the experiment was initiated, some incoming larvae could still establish, since the number of inhibited larvae recovered from calves which received no additional larvae was half that recovered from calves which continued to receive them. Calves, treated so that adult worms were regularly removed, harbored fewer arrested larvae than did their corresponding controls. Michel concluded that the absence of adult worms stimulated the resumption of development among arrested larvae. Although the design of the experiment was such that entering larvae should not have been able to establish, it has already been noted that some larvae continued to be accepted. However, although the results are not inconsistent with Michel's earlier observation that the presence of adult worms increases the number of larvae entering the arrested state, they provide no new evidence in support of this.

Number of Worms in a Single Invasion

Many authors dealing with a variety of gastrointestinal trichostrongyles have found a density-dependent effect on developmental arrest, i.e. the size of the dose influences the proportion of larvae becoming arrested. Genera wherein an effect has been reported include: *Ostertagia, Cooperia, Nematodirus, Graphidium, Obeliscoides* and *Haemonchus*. Some of the relevant literature has been cited already in the discussion of immunological induction of arrest. As developmental inhibition has traditionally been considered as one of the manifestations of host resistance, most authors who have been directly interested in the subject has been concerned with immunological aspects of parasitism. However, to the author's knowledge, there is no conclusive evidence that the size of dose acting indirectly via the host's response determines the fraction of the dose entering arrest. The belief that when a large fraction of a large dose arrests, it is an immunologically mediated phenomenon, is based largely on the idea that the greater the dose, the greater the antigenic stimulation and the greater the host's response. In the case of species which do not arrest until the late fourth or early fifth stage, this presents no special conceptual difficulty. However, some species arrest as early third stage larvae, exhibiting no increase in size over that observed upon reaching infectivity as free-living organisms. Dose-dependent effects, if they do occur among such species, would have to be triggered immediately. Thus, for these species at least, an indirect dose-related control expressed via the host's immunological system seems improbable. Presumably there could be a direct effect from the parasite recognizing its own density through some pheromone-like substance.

Nevertheless, many authors have reported that the percentage of the worm burden which arrests varies directly with the size of the dose. Dunsmore (1960), for example, demonstrated such a relationship between arrest and dose in the case of *Ostertagia*. Table III

TABLE III.

The effect of the size of dose on development arrest in Ostertagia *spp. of sheep (from data presented by Dunsmore, 1960).*

Dose	Sheep	Days to slaughter	Percent Recovered	Percent arrested	
				mean	range
1,000	3	14	92.6 - 100.0	1.5	1.1 - 1.9
100,000	3	14	27.1 - 56.7	59.1	35.9 - 75.8

shows that when a single dose of 1,000 infective larvae was given to lambs, a low mean percentage inhibition ($\bar{x}$ = 1.5%) occurred, whereas when 100,000 larvae were given, a high percentage arrested ($\bar{x}$ = 59.1%). The frequency distributions of the length of the worms given at the two doses were strikingly different. At 1,000 larvae per lamb, there was essentially a normal distribution, with most worms measuring from 6-11 mm. In sharp contrast, the frequency distribution of worms recovered from lambs given the high dose was bimodal; 35-75% of the worms measured only 1-2mm, while a second peak in frequency occurred between 5 and 8 mm. Not only were most worms arrested, but those which developed were stunted.

The effect of dose on arrest has been investigated most comprehensively in *Obeliscoides cuniculi* in the rabbit (Russell, Baker, and Raizes, 1966; Hutchinson, Lee, and Fernando, 1972; Michel, Lancaster, and Hong, 1975b). Russell and colleagues infected seven groups of five rabbits with progressively larger doses in the range of 2,500, to 25,000 larvae per kg body weight. Fifty-one days after infection, all rabbits were sacrificed and the worms present were counted differentially. The number of arrested larvae varied directly with the size of the dose, and the authors attributed this relationship to either a "biomass threshold" or an "immunogenic threshold", which controlled the size of the adult population.

Hutchinson, Lee, and Fernando (1972) having shown that low temperature and changes of temperature would induce arrest in *O. cuniculi*, examined the relationship between dose and arrest in cold-conditioned larvae. They infected three groups of five rabbits with 5,000, 10,000 and 15,000 larvae per rabbit and sacrificed the animals after 14 days. They found no correlation between dose and the population inhibited when this was based on the dose given, but they did find a significant correlation when the proportion was based on numbers of worms recovered. They imply that the apparent correlation between size of dose and larval arrest is due to the differential loss of adult worms between infection and post-mortem examination. Michel (1974), in his review of developmental arrest, also takes this position. However, the data presented by Russell,

Baker, and Raizes (1966) show a clear relationship between dose and the absolute number of larvae arrested and, contrary to Hutchinson, Lee, and Fernando (1972) if percentage arrest is recalculated on the original dose, the data of Russell, Baker, and Raizes (1966) do demonstrate a relationship between dosage and arrest. As shown in Table IV, percentage arrest increases with

TABLE IV.

The effect of dose on developmental arrest of Obeliscoides cuniculi *in rabbits (based on data presented by Russell, et al., 1966).*

Dose	No. of larvae recovered	% of dose recovered arrested
2,500	159	6.4
5,000	875	17.5
10,000	2340	23.0
15,000	3042	20.3
20,000	5133	25.5
25,000	7170	28.7

dose from 6.4% (2500 L_3/kg) to 28.7% (25,000 L_3/kg).

Subsequently, in a comprehensive, well-conceived series of experiments, Michel, Lancaster, and Hong (1975b) re-examined the relationship between dose and developmental arrest in *O. cuniculi*. In the first experiment, they infected six groups of six rabbits with doses forming a series between 2,000 and 30,000 larvae on a single occasion, and all rabbits were sacrificed after a single time interval (17 days). This experiment demonstrated clearly that there was a small dose-dependent effect on arrest. At the lowest dose, 1.17% of the dose was recovered inhibited, while at the highest dose, 5.42% of the dose was still in the fourth stage. The latter represented 23.7% of the total worm burden. While it was apparent that, at the higher doses, developing worms were lost more rapidly than arrested worms, it was also obvious that more larvae must have been arrested.

In their second experiment, three groups of 12 rabbits were infected with either 2,000, 10,000 or 30,000 larvae on a single occasion; thereafter four rabbits per group were killed on day five, day 17 or day 49. On day five, a similar proportion of the dose was recovered from each of the groups, indicating that there was no density-dependent initial failure to establish. When the percentage arrest was expressed on the basis of worms which had established (i.e. worms recovered on day five), a definite density-dependent trend emerged, with 0.15, 1.80, and 6.36% arrest corresponding to 2,000, 10,000 and 30,000 larvae/dose, respectively. By the 49th day, this trend was no longer clearly apparent. Between day 17 and day 49, a sharp decrease in arrested larvae occurred, in part at least due to resumed development, which presumably obscured the pattern. In a third experiment, cold conditioned larvae, expected to have a high potential to arrest, were used in an experiment which otherwise was identical to the second in design. On day 17, the proportion of the worms arrested was much greater than in the two previous experiments, and was no doubt due to the chilling of the infective larvae. As in the second experiment, mean percentage arrest, based on worms recovered on day five, increased with the size of the dose: 8.5, 14.6 and 33.8%, respectively, for the increasing series of doses. Michel and co-workers concluded that the larvae respond to both environmental conditioning and density. This experiment will be discussed further in the next section.

INTERACTIONS

In the introduction to this paper, the known stimuli for arrested development were grouped under three main headings, namely parasite-related, environmental, and host-related. These main groupings were presented in a Venn diagram, with its areas of overlap representing the concept that stimuli from among those listed under the different headings, interact either to cause arrest or to increase the proportion of larvae arresting over that which would occur if the stimuli acted independently.

In the preceding section, an experiment was described in which Michel, Lancaster, and Hong (1975b) conditioned larvae of *Obeliscoides cuniculi* for five weeks at $4^{\circ}C$, and dosed them to rabbits at three different rates. Chilling for five weeks is known to induce arrest. The chilled larvae given to rabbits at the two high dosages showed a greater tendency to arrest than would be expected were the two factors, dose and microclimate, acting independently. This is precisely the kind of interaction which, in the author's view, might be expected.

If one considers density a parasite-related factor, then the example just discussed demonstrates the occurrence of an interaction between environmental and parasite-related stimuli for arrest. In the same class of interactions, one can include the obvious interaction between strain, i.e. another parasite-related factor, and the ability to respond to environmental factors. The possible interaction between the presence of adult worms and environmental conditioning of larvae apparently has not been investigated.

With reference to interactions between host factors such as age, sex and species, and parasite factors such as strain, size of dose and presence of adult worms inducing arrest, no direct published information exists. However, Madsen (1962) has theorized that inhibited development is the result of the interplay between such factors. It seems probable that he will be proven correct, since two of the salient features of developmental arrest are that it is rarely complete, i.e. only a fraction of the population arrests, and that it is facultative. This suggests complex causation.

Furthermore, it has been the general experience of investigators in this field that there is marked individual variation in the number of larvae found arrested between hosts infected from a single pool of larvae. Indeed, idiosyncratic behavior with regard to failure to arrest has often resulted in the premature death of experimental animals. Again, interactions between host factors and parasite factors as determinants of arrest are suggested.

There is little information from experiments designed specifically to investigate the interaction between host factors (sex, age and innate or acquired resistance) and environmental factors (chilling, photoperiod, etc.). Indeed, since environmental stimuli for arrest have been recognized and come under study only recently, investigators have generally been concerned with holding other variables steady. However, at least one experiment of some relevance was conducted in the period before it was known that external environmental factors could act as arrest-provoking stimuli, the arrested state being considered primarily, if not exclusively, a manifestation of host resistance. In an investigation designed to test the duration of acquired immunity to the lungworm, *Dictyocaulus viviparus*, the resistance of cattle was challenged at intervals up to 27 months after vaccination (Michel, et al., 1965), and season (i.e. an unidentified environmental factor) appears to have emerged as an important confounding variable (Michel, 1974). The original paper gives differential counts for adult and larval worms for individual hosts, and it is possible to examine the data for an interaction between immune status and season with regard to the induction of arrest. That *D. viviparus* normally overwinters in the host as an inhibited immature adult is well known from other investigations (Gibbs, 1973; Gupta and Gibbs, 1975; Supperer and Pfeiffer, 1971).

After dividing 90 calves into three lots of 30, Michel and his colleagues immunized one lot with X-irradiated larvae, another with normal larvae and kept the third untreated to use as challenge controls. Six calves from each of these groups were challenged at either 3, 6, 12, 18 or 27 months after vaccination. At each of these intervals, the challenge infections were allowed to develop in half of the cattle for 10 days and in the remainder for 30 days, when they were killed and their worms differentially counted.

Calves of the uninfected control group, when challenged after 12 or 27 months and examined at necropsy at 30 days, harbored parasite burdens with a remarkable representation of arrested worms (54.9 and 56.8%, respectively). Michel (1974), commenting retrospectively, says "On two of the five occasions on which the cattle of Michel, et al., were challenged, over 50% of the worms recovered from susceptible controls after 30 days were under 3 mm in length; on the other three occasions the number of arrested worms was negligible. There was no obvious feature common to the two occasions and absent from the other three unless it be that the two occasions were in autumn and the other three in spring".

Among immunized animals on these same two occasions, the mean percentage of the worm burden recovered arrested after 30 days was 83.5% and 68.5% in cattle immunized with X-irradiated larvae, and 98.4% and 94.6% in those vaccinated with normal larvae, suggesting that the number arrested may have been enhanced by an interaction between environmental and host factors. However, critical examination of the data indicates that a more complicated explanation is required.

Acquired immunity does play a role in the induction of developmental arrest in *D. viviparus* (Michel, 1974). Examination of the data on the three occasions when autumn was not a confounding factor shows that the mean percentage of worms arrested at 30 days, whether calculated on the basis of dose, on worms establishing initially (i.e. 10 day necropsies), or on the final worm burden, was greater in the immunized groups than in the previously uninfected controls.

Thus, there is reasonably good evidence that either acquired resistance or an unidentified environmental factor, can, while acting alone, signal larvae to arrest; it would seem probable, then, that the effects of immunity and an arrest-provoking environment should be additive at least. However, this is not demonstrable when percentage arrest occurring in the fall is calculated on the basis of dose or on the basis of worms establishing initially. Instead, the percentage arrest found to occur in immunized hosts when these are examined 30 days after challenge is less than in normal controls. It is only when arrest is expressed as a percentage of worms recovered at necropsy that arrest maximizes in the immunized groups. Differential loss of worms accounts for the high percentage arrest in immunized hosts, and by the thirtieth day after challenge, any interaction or additive effect of acquired resistance on the environmental induction of arrest has been masked by the rapid loss of worms from immunized animals. This is reminiscent of our own experience with *Ancylostoma caninum* in dogs.

In highly susceptible young pups following infection by stomach tube, the larvae of *Ancylostoma caninum* normally develop directly to adulthood without leaving the small intestine, i.e. there is no somatic migration and no larval arrest in the somatic musculature (see Miller, 1971, for references). However, a small percentage of larvae become arrested in the intestine, and attempts to induce high levels of developmental inhibition by various environmental manipulations involving short photoperiod, conditioning at low constant temperature, day-night variation at autumnal temperature and exposure to sudden change in temperature have been reported in abstract (Schad, 1974). The most successful treatment has been a sudden chilling; this, coupled with a sharp reduction in infectivity, yields worm burdens in which 60-70% of the parasites (and sometimes even higher percentages) are arrested.

To determine whether, in this host-parasite system, there is an interaction between the immunological status of the host and the expression of an environmentally induced potential to arrest, Schad, Stromberg, and Weiner (unpublished) infected young pups, reared helminth-naive, as shown in Table V. If the ability to

TABLE V.

Experimental design for an investigation of the relationship between immunological status and arrested development of Ancylostoma caninum *in pups.*

Treatments	GROUP A: Immunized	GROUP B: Naive	GROUP C: Immuno-Suppressed	GROUP D:[1] Vaccine Controls
X-irradiated larval vaccine	+	-	-	+
Antilymphocyte serum (ALS)	-	-	+	-
Chilled hookworm larvae[2]	+	+	+	-

[1] Group D served to determine whether larvae from the vaccine occurred in the gut, and would have to be taken into account in assessing Group A.
[2] Larvae having a high potential for developmental arrest.

arrest is completely determined by a physiological change in the larvae induced by chilling, then it was expected that arrest would be approximately equal across groups receiving conditioned larvae. If, on the other hand, the host's immune response is involved, it was expected that arrest would be greatest in the vaccinated pups, intermediate in the naive pups and least in those immunosuppressed.

Three groups of four littermates were assigned to this experiment so that one pup from each litter occurred in each treatment group; two pups were held as reserves should death(s) occur in the immunosuppressed group. Two pups from each litter were vaccinated with X-irradiated larvae; one, instead of the two recommended vaccinations, was given, as it was conceivable that double vaccination would cause expulsion of arrested larvae, thereby masking any additive effect on arrest. Three pups, one from each litter, were given antilymphocytic serum (ALS) beginning eight days before infection and periodically thereafter, but one died prematurely. Its substitute received three daily injections of ALS by the day of infection when, as judged by lymphocyte counts, a degree of immunosuppression equivalent to that observed in the original group had occurred. The successful maintenance of immunosuppression was monitored throughout the experiment by lymphocyte counts and other standard immunological techniques. These showed that suppression had been maintained. Twenty-seven days after vaccination, larvae having a high potential for arrest were administered at the rate of 200/lb by stomach tube to Groups A, B and C. The pups were sacrificed three weeks after infection.

Table VI indicates that immunosuppression significantly affected all parasitological measures of infection. In group C, a greater percentage of the dose established, matured or became arrested than in any of the other groups. In terms of worm burden, a significantly greater proportion of the worms recovered were arrested in the suppressed than in the naive pups, but no statistically significant difference was demonstrated between the suppressed and the vaccinated groups. These results proved contrary to both alternative patterns of between-group parasitism predicted by the hypotheses under investigation, i.e., that minimal, rather than maximal arrest would occur in immunosuppressed animals, or that no difference in arrest would be observed between groups.

The most obvious explanation of these results

TABLE VI.

Arrested development of parasitic third stage Ancylostoma caninum larvae in pups of differing immunological status.

Treatment Group	Dog No.[1]	Worms Recovered		% based on dose:[2]			% of burden Arrested[2]
		Adult	Larvae	Adults	Larvae	Total	
A. Immunized	ED4	25	33	2.2	3.0	5.2	56.9
	EL3	13	79	2.0	12.2	14.2	85.9
	EZ3	14	162	1.3	14.8	16.1	92.0
				$\bar{x}$ 1.9±0.5	$\bar{x}$ 10.0±6.2	$\bar{x}$ 11.8±5.8	$\bar{x}$ 78.3±18.8
B. Naive	ED1	26	46	1.8	3.1	4.8	63.9
	EL2	36	91	3.7	9.3	13.0	71.7
	EZ2	18	101	1.4	7.8	9.3	84.9
				$\bar{x}$ 2.3±1.2	$\bar{x}$ 6.7±3.2	$\bar{x}$ 9.0±4.1	$\bar{x}$ 73.5±10.6
C. Immuno- suppressed	ED2	45	176	3.7	14.3	18.0	79.6
	EV1	44	214	5.0	25.5	29.5	83.0
	EZ1	36	408	2.9	33.0	35.9	91.9
				$\bar{x}$ 3.8±1.1	$\bar{x}$ 23.9±9.4	$\bar{x}$ 27.8±9.1	$\bar{x}$ 84.8±5.6
Statistical Comparisons[3,4]	A vs. B			N.S.	N.S.	N.S.	N.S.
	B vs. C			< .05	< .05	< .05	< .02
	A vs. C			< .05	< .01	< .01	N.S.

[1] Dogs identified by similar combinations of letters were littermates.
[2] Percentages and means ± standard deviations before arcsine transformation.
[3] N.S. = not significant.
[4] Probabilities determined by paired comparison t-test on arcsine transformed data.

depends on differential retention of worms. Apparently, a proportion of the infective larvae were programmed for arrest, and immunosuppression did not affect this proportion, i.e., the suppressed animals retained both those larvae destined for arrest and those capable of immediate development more readily than did either the naive or vaccinated pups. In Groups A and B, parasites were eliminated differentially causing a shift in the percentage arrested.

If immune status is relevant, then why did the vaccinated group fail to differ from the suppressed with regard to the fraction of the worm burden arrested and, in fact, why did the vaccinates not differ from the naive group in any of the parasitological criteria examined? Possibly our vaccinating procedure was too conservative; only one of the two prescribed doses was used to minimize the risk that a strongly immunized host would eliminate arrested larvae completely. Thus, functionally, the immune status of Group A might not have differed effectively from Group B. However, more critical examination of the data suggests that acquired immunity may contribute to the maximal expression of a potential to arrest. If pup ED4 is excluded from the vaccinated group, the percentage is as great as occurred in the suppressed group, and the overall pattern which emerges is that maximal arrest is observed in the polar groups, i.e., immunized and suppressed (Table VI). Consideration of the data without this pup is justified, since, at necropsy, lengths of shed intestinal epithelium were recovered from its feces. This complication could have resulted in the loss of worms, particularly larvae. If this suggestion has merit, the full understanding of arrested development and survival in the arrested state will be complex. It will involve a proportion of infective larvae having a potential for arrest, the full expression of which is determined by the host's immune response, and after the larvae have become arrested, a facet of this response will determine the number retained.

THE PHYSIOLOGICAL BASIS OF ARREST

Periods of developmental arrest are inherent in the

basic pattern of nematode growth and development. Thus, as Rogers and Sommerville (1963) have stressed, an interval of quiescence is often associated with ecdysis and, particularly upon reaching the infective stage, this period of arrested development may be prolonged pending entry into a host (Rogers, 1961; Rogers and Sommerville, 1963). When a prolonged period of arrest occurs during inhibited development within a host, it, too, is generally associated with the moult and may be viewed as an extended lethargus (Rogers and Sommerville, 1969). Figure 6 presents Sommerville's well known

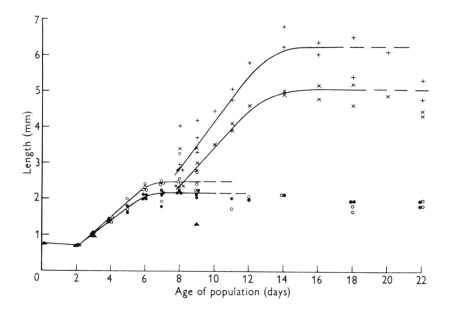

Figure 6. Growth curve of Cooperia curticei *in sheep. Open circles = fourth stage females, solid circles fourth stage males; + = adult males; x = adult females; triangles = sexes not differentiated. Reproduced from Sommerville (1960).*

growth curve of *Cooperia curticei*, which illustrates this concept particularly well. Even when *C. curticei* is developing within a normal time frame, its growth is interrupted briefly as it begins the fourth moult. This results in a short plateau. If larvae become inhibited in their development, growth is interrupted at precisely this point, as may be seen in the figure.

Rogers and his colleagues, as well as other workers who were influenced by his work and ideas, have focused mainly on the triggering of ecdysis and the sequence of events leading to the production of exsheathing fluid. This they postulated to be a multiphasic process involving a signal from the host, the production of neuroendocrine substances with an attendant release of lytic enzymes facilitating the rupture of the cuticle. Subsequent investigations have supported this concept (Davey, 1970). Less consideration appears to have been given to the possibility that a sequence of similar steps is required to initiate resumption of growth once the moult has been completed. However, Samoiloff (1973) has recently shown by laser microbeam surgery that, while ecdysis is regulated by organ systems in the area of the nerve ring, general growth is under the control of systems located in the region of the hind-gut. The first of these observations is in agreement with what was already known about moulting, but the second is new and of particular interest in relation to developmental arrest of parasitic stages. Many nematodes complete an ecdysis before they become arrested in their hosts. By distinguishing clearly between failure to complete ecdysis and failure to maintain growth, it remains possible to consider arrested development occurring under a variety of circumstances as an extended period of quiescence requiring some triggering mechanism for the continuance of development. In this connection, a report of Bruce and Armour (1974) is of great interest. These authors have reported that during arrest of *Ostertagia ostertagi* larvae, a marked accumulation of neurosecretory material occurs, particularly in cells associated with the anal ganglia. Apparently this system is ready to reactivate the larvae when appropriately triggered.

ARRESTED DEVELOPMENT AND THE REGULATION OF HELMINTH POPULATIONS: SYNTHESIS, SPECULATION AND SUMMARY

It has been suggested frequently that, in the evolution of associations between hosts and their metazoan parasites, the latter have evolved so that their presence is not detected readily by the host's defence systems. Among several adaptive strategies which would protect the parasite from immunological attack by the host is evolution toward antigenic similarity with the host, i.e. molecular mimicry (Damian, 1964; Dineen, 1963; Donald, et al., 1964; Schad, 1966; Sprent, 1962). For reasons given by Capron, et al. (1968), the theory of antigenic convergence cannot be valid for parasites markedly lacking in host specificity, since such parasites could not code genetically for the diverse antigens of a broad spectrum of hosts. Other mechanisms which may apply in these situations are a host-induced production of host-like substances by the parasite (Capron, et al., 1968), incorporation or coating of host material onto the surface of the parasite (Smithers, Terry, and Hockley, 1968; 1969) or depression of the host's immune responsiveness by the parasite (Barriga, 1975; Cypress, Lubinecki, and Hammon, 1973; Faubert, 1976). As was indicated previously, one tenet of the several theories based on antigenic convergence is that the host fails to react to the presence of the parasite until it attains some critical biomass. At this threshold level, antigenic stimulation becomes sufficient to elicit an immune response which limits the parasite's biomass and its ability to increase its abundance further. Among the several possible manifestations of host resistance observed when the parasite biomass exceeds the threshold is developmental arrest. Thus, in the well-adapted relationship, worms are stored by the host as innocuous, dormant larvae before parasite density exceeds dangerous levels. Clearly, the significance of arrest is seen in terms of population regulation.

Much of the research on arrested development and population regulation has been done by veterinary parasitologists concerned with the parasites of grazing animals. These workers have not been particularly concerned with relatively benign, well-adapted parasites, but rather with the more pathogenic parasites of livestock whose epidemiology and control they have sought to understand. They, too, have usually viewed inhibited development as a manifestation of host resistance, particularly in the repeatedly challenged host. This concept has been developed most completely by Soulsby

(1958; 1966) with regard to the epidemiology of gastrointestinal parasitism in sheep in areas having a well defined winter which interrupts transmission and where the so-called "spring rise" in nematode egg counts in ewes is an important source of infection for lambs. According to his theory, as spring and summer progress, the host, after repeated exposures to infection, manifests an acquired immunity which causes larvae newly acquired in autumn to arrest. Subsequently, without further antigenic stimulation during the winter, and with a relaxation of responsiveness during late pregnancy and lactation, resistance in ewes wanes, permitting resumption of larval development in the spring. Clearly this series of events would prevent the excessive accumulation of adult worms toward the end of the transmission season. It would also synchronize events in the life of the parasite with those occurring in the host, and thereby facilitate transmission.

Gordon (1973), Michel (1974) and Schad, et al. (1973) have recently viewed arrested development from a more parasite-centered standpoint. Gordon believes that hypobiosis (arrested development) is an adaptation of the parasite to survive periods of adversity, both external (climatic) and internal (host resistance), dormant worms being more resistant to adverse environments than those that actively develop. He considers developmental arrest a special phenomenon, perhaps a special adaptation on the part of the gastrointestinal nematodes of grazing animals, "to counter immunogenic attack which may be a consequence of the increased exposure of host to parasite in modern husbandry". He emphasizes in particular the malfunctions in the system. He sees hypobiosis as enhancing the potential for outbreaks of parasitic disease through the accumulation of arrested worms, "followed by mass development".

Michel and his colleagues have made numerous important contributions to knowledge of developmental arrest in nematodes (references in Michel, 1974; 1976). In his reviews of the subject (1968; 1974), he considers arrest to be an adaptation for synchronizing events in the life history of the parasite with events occurring either within its host, or in the external environment. Defining punctuality as being at the right place at the right time, Michel (1974) states that "punctuality...is the essence of successful parasitism and the most effective aid to punctuality is the ability to mark time". He rejects the idea that arrest plays a role in regulating parasite populations (p. 339); this is surprising since he also states that, with regard to arrest, whether induced by the immune response of the host or by size of dose, an "irreducible core of evidence remains" that these mechanisms do function and that feedback mechanisms controlling resumption of development do exist.

Schad, et al. (1973), in a paper concerned largely with arrest in human hookworms, stressed the fact that selection would operate strongly against a parasite that wasted energy seeding an inhospitable external environment with its eggs. However, it is evident that, in this instance as in most others where developmental arrest is seasonal, arrest will serve two functions. It will synchronize the life cycle of the parasite with the external changes in climate, and it will serve to regulate the worm population.

In West Bengal, larval hookworm populations in the external environment increase in the course of the monsoon season and maximize at its end. The intensity of transmission must therefore increase during the rainy season and there is evidence that it does (Nawalinski, 1974). If all the larvae that penetrated successfully at the end of the monsoon matured promptly, worm burdens would peak sharply at the end of the wet season. Instead, they climb dramatically before the monsoon begins, due to the relatively synchronous maturation of larvae which did not develop in the previous year. Thus, instead of adult populations rising slowly at the beginning of the monsoon and peaking sharply at its end, a more asymptotic situation occurs, with a broad, truncated peak coinciding with the entire wet season (Figure 7).

Viewed in terms of _all_ _stages_ in the life cycle, populations of both the free-living stages and of adult worms will begin to decrease soon after the close of the rainy season, while populations of arrested larvae in man will have begun their increase. The overall effect will be that changes in the number of arrested larvae will dampen out the oscillations in hookworm abundance. Thus, although Schad, et al. (1973) suggested that

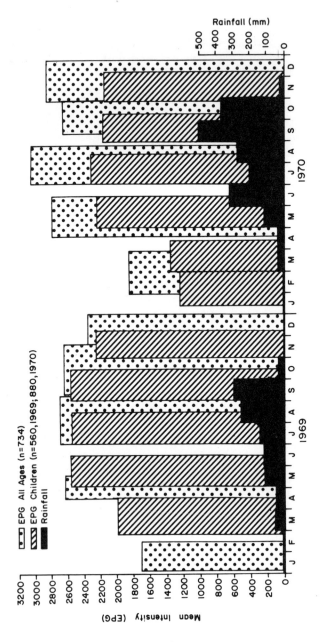

Figure 7. Seasonal variation of hookworm egg counts of two different groups of rural Bengalis.

arrest in *A. duodenale* is an adaptation to a seasonally unfavorable external environment, it is apparent that it also plays an important part in regulating the biomass of the parasite.

With regard to the Bengal strain of *A. duodenale*, it is particularly remarkable that, in two self-induced infections of man, absolute arrest occurred, i.e. 100% of the worms which penetrated arrested. As was mentioned previously, this led Nawalinski and Schad (1974) to suggest that our particular strain might have an obligatory, genetically determined, long prepatent period.

Waller and Thomas (1975) have suggested that some geographic races of parasites occurring in areas where unfavorable seasons are particularly long and harsh may have evolved so that arrest is no longer facultative, and, hence, but one generation per year occurs. Although Waller and Thomas may not have completely excluded the possibility that arrest is induced seasonally by environmental factors, their concept is interesting and, by analogy with the ways in which insects solve the same problems, it is not totally unexpected. In these peripheral parts of the range of a parasite, where seasonally the climate interrupts transmission for most of the year, one might expect parasites to be 'r' selected (see Esch, Hazen, and Aho, this volume), and one correlate of strong 'r' selection might well be obligatory arrest, or at least a particularly efficient system whereby seasonal arrest is induced by a signal, environmental or otherwise, which reliably predicts the coming of an unfavorable season.

Evidence for this may be that in Canada, where *Haemonchus* does not overwinter successfully as a free-living organism, a short photoperiod rather than low temperature is the signal for arrest. The former is the more reliable of these signals, and 90-100% of larvae acquired under field conditions arrest (Blitz and Gibbs, 1972a; Gibbs, 1973). In another peripheral part of its range (northeast England), where cool summers offer less opportunity for more than one generation of free-living larvae to occur per year, Waller and Thomas (1975) suggest that arrest in the host has been assured by incorporating obligatory dormancy into the fundamental life history plan.

According to ecological theory, in more benign

climates the parasite should be more 'K' selected. In this connection, it is of interest that haemonchosis in sheep in the area of Ithaca, New York is an "orderly disease", and that, in most sheep, parasite populations are well regulated and "self-cure" does not occur (Whitlock, 1966). Further south in the area of Beltsville, Maryland, only few arrested larvae occur in sheep during winter, and "spring rise" is unknown (Kates, personal communication).

Gordon has emphasized that seasonal arrest with the concommitant massive accumulation of larvae carries the paradoxical risk that it enhances the potential for an outbreak. Indeed, instability can ensue if resumption of development is particularly synchronous, and if developing and/or adult worms are not expelled efficiently. However, although dense populations of arrested larvae commonly occur in grazing animals during late fall and winter, only few individuals are clinically affected by the resumption of larval development. Furthermore, in a large scale investigation of resumption of development in *O. ostertagi*, Michel, Lancaster, and Hong (1976) demonstrated that reactivation was not as synchronous as had been suggested previously. Some larvae resumed development throughout the winter, but when larvae had not developed by spring, they did so in a short space of time. Even then, however, cattle did not suffer clinically. Clinical grades of infection, attributable to resumption of larval development, were observed earlier and occurred in only few animals; these had never been robust and did not react as did the others to limit their adult worm burdens to tolerable levels. Invariably the survivors showed large populations of arrested larvae, whereas those that died did not.

Apparently, in these severely affected individuals, larvae which would have normally remained arrested suddenly developed *en masse*. Not only is this another reason for thinking that host resistance is an important factor in the maintenance of arrest, but, since it was only the occasional host which failed to regulate its parasite population, even under the new and unnatural conditions of husbandry, it appears that storage of parasites as dormant larvae will curb outbreaks more often than set the stage for them.

In part at least, the so-called "spring rise" in nematode egg counts in herbivores depends on the seasonal maturation of arrested larvae (Blitz and Gibbs, 1972b). For its complete expression, it requires what Gordon (1973) has termed the "post-parturient relaxation of resistance" and clearly it enhances transmission of parasites to new crops of young susceptible hosts. In carnivores the reactivation of larval worms at about the time of parturition and for some time during lactation is a similar phenomenon leading to vertical transmission. These sequences of events do indeed synchronize events in the life of the parasite with those occurring in the life of the host. The question arises then: Are these instances where the regulatory aspect of arrest becomes unimportant? Initially it appears obvious that the significance of these events is transmission of parasites into new groups of susceptibles. However, this may be entirely too simple; the transmission of parasites to the neonate probably has significance for the population ecology of host-parasite systems which is barely perceived at this time. This is suggested by the observations by Kassai (1967), Kassai and Aitken (1967) and Dineen and Kelly (1973), who found that if neonatal rats were infected with *Nippostrongylus brasiliensis*, the usual rapid expulsion of the worms after two weeks did not occur. Instead, these rats harbored infections which persisted into adulthood, and under some circumstances survived challenge infections which were rejected. Apparently the timing of first exposure to infection can markedly effect the long term stability of parasite populations, and, this, in turn, would have profound effects on the population ecology of host-parasite associations.

In summary, then, developmental arrest in nematodes during some stage in their parasitic development rarely, if ever, has a single function. It is an oversimplification to consider the significance of arrest only in terms of synchronizing the life cycle of the parasite with changing external environmental conditions, or with events in the life of the host. Storage of larvae in the host in a quiescent form limits large oscillations in parasite abundance or biomass, and thus arrest has a regulatory function. Large oscillations place the host, the parasite, or both, in jeopardy, and strong selection for mechanisms to dampen these oscillations would be expected. Arrested development is such a mechanism.

ACKNOWLEDGEMENTS

This manuscript was supported by United States Public Health Service Grant AI 11852.

LITERATURE CITED

1. ANDERSON, N. 1972. Ostertagiasis in beef cattle. *Vict. Vet. Proc. 30:* 36-38.
2. ANDERSON, N., J. ARMOUR, F. W. JENNINGS, J. S. D. RITCHIE, and G. M. URQUHART. 1965. Inhibited development of *Ostertagia ostertagi*. *Vet. Rec. 77:* 146-147.
3. ARMOUR, J. 1970. Bovine ostertagiasis: a review. *Vet. Rec. 86:* 184-190.
4. ARMOUR, J. and R. G. BRUCE. 1974. Inhibited development in *Ostertagia ostertagi* infections - a diapause phenomenon in a nematode. *Parasitology 69:* 161-174.
5. ARMOUR, J., F. W. JENNINGS, and G. M. URQUHART. 1969a. Inhibition of *Ostertagia ostertagi* at the early fourth larval stage. I. The seasonal incidence. *Res. Vet. Sci. 10:* 232-237.
6. ARMOUR, J., F. W. JENNINGS, and G. M. URQUHART. 1969b. Inhibition of *Ostertagia ostertagi* at the early fourth larval stage. II. The influence of environment on host or parasite. *Res. Vet. Sci. 10:* 238-244.
7. AYALEW, L., J. L. FRECHETTE, R. MALO, and C. BEAUREGARD. 1973. Gastrointestinal nematode populations in stabled ewes of Rimouski region. *Can. J. Comp. Med. 37:* 356-361.
8. AYALEW, L. and H. C. GIBBS. 1973. Seasonal fluctuations of nematode populations in breeding ewes and lambs. *Can. J. Comp. Med 37:* 79-89.
9. BANKS, A. W. and J. F. MICHEL. 1960. A controlled trial of o, o-dimethyl 2, 2, 2 trichloro-1-hydroxymethyl phophonate as an anthelimintic

against *Ostertagia ostertagi* in calves. *Vet. Rec. 72:* 135-136.
10. BARRIGA, O. O. 1975. Selective immunodepression in mice by *Trichinella spiralis* extracts and infections. *Cellular Immunol. 17:* 306-309.
11. BENZ, G. W. and A. C. TODD. 1969. Observations on the organization of nematode developmental stages present in some natural and experimental infections. *Trans. Amer. Microsc. Soc. 88:* 89-94.
12. BINDERNAGEL, J. A. and A. C. TODD. 1972. The population dynamics of *Ashworthius lerouxi* (Nematoda:Trichostrongylidae) in African buffalo in Uganda. *Brit. Vet. J. 128:* 452-455.
13. BLITZ, N. M. and H. C. GIBBS. 1971. Morphological characterization of the stage of arrested development of *Haemonchus contortus* in sheep. *Can. J. Zool. 49:* 991-995.
14. BLITZ, N. M. and H. C. GIBBS. 1972a. Studies on the arrested development of *Haemonchus contortus* in sheep. I. The induction of arrested development. *Internat. J. Parasitol. 2:* 5-12.
15. BLITZ, N. M. and H. C. GIBBS. 1972b. Studies on the arrested development of *Haemonchus contortus* in sheep. II. Termination of arrested development and the spring rise phenomenon. *Internat. J. Parasitol. 2:* 13-22.
16. BRUCE, R. G. and J. ARMOUR. 1974. Ultrastructural changes in *Ostertagia ostertagi* associated with inhibition. *Proc. Third Internat. Cong. Parasitol. 1:* 452.
17. BRUNSDON, R. V. 1973. Inhibited development of *Haemonchus contortus* in naturally acquired infections in sheep. *New Zeal. Vet. J. 21:* 125-126.
18. CAPRON, A., J. BIGUET, A. VERNES, and D. AFCHAIN. 1968. Structure antigenique des helminthes. Aspects immunologiques des relations hote-parasite. *Pathol. Biol. 16:* 121-138.
19. CONNAN, R. M. 1971a. *Hyostrongylus rubidus:* The size and structure of worm populations in adult pigs. *Vet. Rec. 89:* 186-191.

20. CONNAN, R. M. 1971b. The seasonal incidence of inhibition of development in *Haemonchus contortus*. *Res. Vet. Sci. 12:* 272-274.
21. CONNAN, R. M. 1974. Arrested development in *Chabertia ovina*. *Res. Vet. Sci. 16:* 240-243.
22. CONNAN, R. M. 1975. Inhibited development in *Haemonchus contortus*. *Parasitology 71:* 239-246.
23. CYPRESS, R. H., A. S. LUBINECKI, and W. M. HAMMON. 1973. Immunosuppression and increased susceptibility to Japanese B encephalitis virus in *T. spiralis* infected mice. *Proc. Soc. Exp. Biol. 143:* 469.
24. DAMIAN, R. T. 1964. Molecular mimicry: antigen sharing by parasite and host and its consequences. *Amer. Nat. 98:* 129-149.
25. DAVEY, K. G. 1970. Hormones, environment and development in nematodes. *In:* H. van den Bossche (ed.), Comparative Biochemistry of Parasites. Academic Press, New York.
26. DINEEN, J. K., A. D. DONALD, B. M. WAGLAND, and J. OFFNER. 1965. The dynamics of the host-parasite relationship. III. The response of sheep to primary infection with *Haemonchus contortus*. *Parasitology 55:* 515-525.
27. DINEEN, J. K. 1963. Immunological aspects of parasitism. *Nature, Lond. 197:* 268-269.
28. DINEEN, J. K., A. D. DONALD, B. M. WAGLAND, and J. H. TURNER. 1965. The dynamics of the host-parasite relationship. II. The response of sheep to primary and secondary infection with *Nematodirus spathiger*. *Parasitology 55:* 163-171.
29. DINEEN, J. K. and J. D. KELLY. 1973. Immunological unresponsiveness of neonatal rats to infection with *Nippostrongylus brasiliensis:* The competence of neonatal lymphoid cells in worm expulsion. *Immunology 25:* 141-150.
30. DONALD, A. D., J. K. DINEEN, J. H. TURNER, and B. M. WAGLAND. 1964. The dynamics of the host-parasite relationship. I. *Nematodirus spathiger* infection in sheep. *Parasitology 54:* 527-544.
31. DUNSMORE, J. D. 1960. Retarded development of

of *Ostertagia* species in sheep. *Nature, Lond. 186:* 986-987.
32. DUNSMORE, J. D. 1961. Effect of whole-body irradiation and cortisone on the development of *Ostertagia* spp. in sheep. *Nature, Lond. 4798:* 139-140.
33. DUNSMORE, J. D. 1963. Effect of the removal of an adult population of *Ostertagia* from sheep on concurrently existing arrested larvae. *Aust. Vet. J. 39:* 459-463.
34. EHRENFORD, F. A. 1956. Canine ascariasis - a potential zoonosis. *J. Parasitol. 42 (suppl.):* 12-13.
35. FAUBERT, G. M. 1976. Depression of plaque-forming cells to sheep red blood cells by the new-born larvae of *Trichinella spiralis*. *Immunology 30:* 485.
36. FERNANDO, M. A., P. H. G. STOCKDALE, and G. C. ASHTON. 1971. Factors contributing to the retardation of development of *Obeliscoides cuniculi* in rabbits. *Parasitology 63:* 21-29.
37. GIBBS, H. C. 1973. Transmission of parasites with reference to the strongyles of domestic sheep and cattle. *Can. J. Zool 51:* 281-289.
38. GIBSON, T. E. 1950. Critical tests of phenothiazine as an anthelmintic for horses. *Vet. Rec. 62:* 341-343.
39. GIBSON, T. E. 1953. The effect of repeated anthelmintic treatment with phenothiazine on the faecal egg counts of houses horses, with some observations on the life cycle of *Trichonema* spp. in the horse. *J. Helminthol. 26:* 29-40.
40. GIBSON, T. E. 1959. The development of resistance by sheep to infection with the nematodes *Nematodirus filicollis* and *Nematodirus battus*. *Brit. Vet. J. 115:* 120-123.
41. GORDON, H. MCL. 1973. Epidemiology and control of gastrointestinal nematodes of ruminants. *Adv. Vet. Sci. Comp. Med. 17:* 395-437.
42. GORDON, H. MCL. 1974. Hypobiosis, haemonchosis and the hytherograph. *Proc. Third Internat. Cong. Parasitol. 2:* 750-751.
43. GREVE, J. H. 1971. Age resistance to *Toxocara*

canis in ascarid-free dogs. *Amer. J. Vet. Res. 32:* 1185-1192.
44. GUPTA, R. P. and H. C. GIBBS. 1975. Infection patterns of *Dictyocaulus viviparus* in calves. *Can. Vet. J. 16:* 102-108.
45. HART, J. A. 1964. Observations on the dry season strongyle infestations of Zebu cattle in Northern Nigeria. *Brit. Vet. J. 120:* 87-95.
46. HERLICH, H. 1967. Effects of *Cooperia pectinata* on calves: two levels of repeated oral inoculation. *Amer. J. Vet. Res. 28:* 71-77.
47. HOTSON, I. K. 1967. Ostertagiosis in cattle. *Aust. Vet. J. 43:* 383-387.
48. HUTCHINSON, G. W., E. H. LEE, and M. A. FERNANDO. 1972. Effects of variations in temperature on infective larvae and their relationship to inhibited development of *Obeliscoides cuniculi* in rabbit. *Parasitology 65:* 333-342.
49. KASSAI, T. 1967. Immunological tolerance to *Nippostrongylus brasiliensis* infection in rats. *In:* E. J. L. Soulsby (ed.), The Reaction of Host to Parasitism. Academic Press, New York.
50. KASSAI, T. and I. D. AITKEN. 1967. Induction of immunological tolerance in rats to *Nippostrongylus brasiliensis* infection. *Parasitology 57:* 403-418.
51. KELLY, J. D. 1973. Immunity and epidemiology of helminthiasis in grazing animals. *New Zeal. J. 21:* 183-194.
52. LEE, K. T., M. D. LITTLE, and P. C. BEAVER. 1975. Habitat of *Ancylostoma caninum* in some mammalian hosts. *J. Parasitol. 61:* 589-598.
53. MADSEN, H. 1962. The so-called tissue phase in nematodes. *J. Helminthol. 36:* 143-148.
54. MALCZEWSKI, A. 1970. Gastro-intestinal helminths of ruminants in Poland. III. Seasonal incidence of the stomach worms in calves, with consideration of the effect of the inhibition phenomenon on the spring rise phenomenon. *Acta Parasit. Polon. 18:* 417-434.
55. MCKENNA, P. B. 1973. The effect of storage on the infectivity and parasitic development of third-stage *Haemonchus contortus* larvae in sheep. *Res. Vet. Sci. 14:* 312-316.

56. MICHEL, J. F. 1952a. Self-cure in infections of *Trichostrongylus retortaeformis* and its causation. *Nature, Lond. 169:* 881.
57. MICHEL, J. F. 1952b. Inhibition of development of *Trichostrongylus retortaeformis*. *Nature, Lond. 169:* 933-934.
58. MICHEL, J. F. 1963. The phenomena of host resistance and the course of infection of *Ostertagia ostertagi* in calves. *Parasitology 53:* 63-84.
59. MICHEL, J. F. 1968. Immunity to helminths associated with the tissues. *In:* A. E. R. Taylor (ed.), Immunity to Parasites (Vol. 6). Symposium Brit. Soc. Parasitol. Blackwell Sci. Pub., Oxford and Edinburgh.
60. MICHEL, J. F. 1970. The regulation of populations of *Ostertagia ostertagi* in calves. *Parasitology 61:* 435-447.
61. MICHEL, J. F. 1971. Adult worms as a factor in the inhibition of development of *Ostertagia ostertagi* in the host. *Internat. J. Parasitol. 1:* 31-36.
62. MICHEL, J. F. 1974. Arrested development of nematodes and some related phenomena. *Adv. Parasitol. 12:* 279-366.
63. MICHEL, J. F. 1976. The epidemiology and control of some nematode infections in grazing animals. *Adv. Parasitol. 14:* 355-397.
64. MICHEL, J. F., M. B. LANCASTER, and C. HONG. 1973. Inhibition of development: variation within a population of *Ostertagia ostertagi*. *J. Comp. Path. 83:* 351-356.
65. MICHEL, J. F., M. B. LANCASTER, and C. HONG. 1974. Studies on arrested development of *Ostertagia ostertagi* and *Cooperia oncophora*. *J. Comp. Path. 84:* 539-554.
66. MICHEL, J. F., M. B. LANCASTER, and C. HONG. 1975a. Arrested development of *Ostertagia ostertagi* and *Cooperia oncophora*; effect of temperature at the free-living third stage. *J. Comp. Path. 85:* 133-138.
67. MICHEL, J. F., M. B. LANCASTER, and C. HONG. 1975b. Arrested development of *Obeliscoides cuniculi*; the effect of size of inoculum. *J. Comp. Path. 85:* 307-315.

68. MICHEL, J. F., M. B. LANCASTER, and C. HONG. 1976. Observations on the resumed development of arrested *Ostertagia ostertagi* in naturally infected yearling cattle. *J. Comp. Path. 86:* 73-80.
69. MICHEL, J. F., A. MACKENZIE, C. D. BRACEWELL, R. L. CORNWELL, J. ELLIOTT, C. NANCY HEBERT, H. H. HOLMAN, and I. J. B. SINCLAIR. 1965. Duration of the acquired resistance of calves to infection with *Dictyocaulus viviparus*. *Res. Vet. Sci. 6:* 344-395.
70. MILLER, T. A. 1965. Influence of age and sex on susceptibility of dogs to primary infection with *Ancylostoma caninum*. *J. Parasitol. 51:* 701-704.
71. MILLER, T. A. 1970. Potential transport hosts in the life cycles of canine and feline hookworms. *J. Parasitol. 56 (Suppl.):* 238.
72. MILLER, T. A. 1971. Vaccination against the canine hookworm diseases. *Adv. Parasitol. 9:* 153-183.
73. MÜLLER, G. L. 1968. The epizootiology of helminth infestation in sheep in southwestern districts of the Cape. *Onderstepoort J. Vet. Res. 35:* 159-194.
74. NAWALINSKI, T. A. 1974. Dynamics of hookworm parasitism in children. D. Sc. Thesis, The Johns Hopkins University, Baltimore.
75. NAWALINSKI, T. A. and G. A. SCHAD. 1974. Arrested development in *Ancylostoma duodenale:* course of a self-induced infection in man. *Amer. J. Trop. Med. Hyg. 23:* 895-898.
76. REID, J. F. S. and J. ARMOUR. 1972. Seasonal fluctuations and inhibited development of gastrointestinal nematodes of sheep. *Res. Vet. Sci. 13:* 225-229.
77. ROBERTS, F. H. S. and R. K. KEITH. 1959. Observations on the effect of treatment with phenothiazine on the development of resistance by calves to infestation with the stomach worm *Haemonchus placei*. *Aust. Vet. J. 35:* 409-414.
78. ROGERS, W. P. 1939. The physiological ageing of ancylostome larvae. *J. Helminthol. 17:* 195-202.
79. ROGERS, W. P. 1961. The Nature of Parasitism: The Relationship of Some Metazoan Parasites to

Their Host. Academic Press, New York and London.
80. ROGERS, W. P. and R. I. SOMMERVILLE. 1963. The infective stage of nematode parasites and its significance in parasitism. *Adv. Parasitol. 1:* 109-177.
81. ROGERS, W. P. and R. I. SOMMERVILLE. 1969. Chemical aspects of growth and development. *In:* M. Thorkin and B. T. Scheer (eds.), Chemical Zoology (Vol. 3). Academic Press, New York and London.
82. ROSS, J. G. 1965. The seasonal incidence of ostertagiasis in cattle in Northern Ireland. *Vet. Rec. 77:* 16-19.
83. RUSSELL, S. W., N. F. BAKER, and G. S. RAIZES. 1966. Experimental *Obeliscoides cuniculi* infections in rabbits: comparison with *Tricostrongylus* and *Ostertagia* infections in cattle and sheep. *Exp. Parasitol. 19:* 163-173.
84. SAMOILOFF, M. R. 1973. Nematode morphogenesis: localization of controlling regions by laser microbeam surgery. *Science 180:* 976-977.
85. SCHAD, G. A. 1966. Immunity competition, and natural regulation of helminth populations. *Amer. Nat. 100:* 359-364.
86. SCHAD, G. A. 1974. Experimentally induced arrested development of the parasitic larvae of hookworms. *Proc. Third Internat. Cong. Parasitol. 2:* 772-773.
87. SCHAD, G. A., A. B. CHOWDHURY, C. G. DEAN, V. K. KOCHAR, T. A. NAWALINSKI, J. THOMAS, and J. A. TONASCIA. 1973. Arrested development in human hookworm infections: adaptation to a seasonally unfavorable external environment. *Science 180:* 502-504.
88. SCHAD, G. A., E. J. L. SOULSBY, A. B. CHOWDHURY, and H. M. GILLES. 1975. Epidemiological and senological studies of hookworm infection in endemic areas in India and West Africa. *In:* Nuclear Techniques in Helminthology Research. Internat. Atomic Energy Agency, Vienna.
89. SCOTT, J. A. 1928. An experimental study of the development of *Ancylostoma caninum* in normal and abnormal hosts. *Amer. J. Hyg. 8:* 158-204.

90. SMITH, H. J. 1974. Inhibited development of *Ostertagia ostertagi*, *Cooperia oncophora*, and *Hematodirus helvetianus* in parasite-free calves grazing fall pastures. *Amer. J. Vet. Res. 35:* 935-938.
91. SMITHERS, S. R., R. T. TERRY and D. J. HOCKLEY. 1968. Do adult schistosomes masquerade as their hosts? *Trans. R. Soc. Trop. Med. Hyg. 62:* 466-467.
92. SMITHERS, S. R., R. T. TERRY, and D. J. HOCKLEY. 1969. Host antigens in schistosomiasis. *Proc. R. Soc. Lond. Ser. B Biol. Sci. 171:* 483-494.
93. SOMMERVILLE, R. I. 1960. The growth of *Cooperia curticei* (Giles, 1892), a nematode parasite of sheep. *Parasitology 50:* 261-267.
94. SOULSBY, E. J. L. 1958. Immunity to helminths. *Vet. Reviews and Annotations 4:* 1-16.
95. SOULSBY, E. J. L. 1966. The mechanisms of immunity to gastrointestinal nematodes. *In:* E. J. L. SOULSBY (ed.), Biology of Parasites. Emphasis on Veterinary Parasites. Academic Press, New York and London.
96. SPRENT, J. F. A. 1958. Observations on the development of *Toxocara canis* (Werner, 1782) in the dog. *Parasitology 48:* 184-209.
97. SPRENT, J. F. A. 1962. Parasitism, immunity and evolution. *In:* G. S. Leeper (ed.), The Evolution of Living Organisms. Melbourne Univ. Press, Melbourne.
98. STOYE, M. 1973. Untersuchungen uber die Moglichkeit pranataler und galaktogener Infektionen mit *Ancylostoma caninum* Ercolani 1859 (Ancylostomidae beim Hund) *Zbl. Vet. Med. B 20:* 1-39.
99. SUPPERER, R. and PFEIFFER, H. 1971. Zur Uberwinterung des Rinderlungenwurmes im Wirtstier, Berl. *Munch. tierarztl. Mschr. 84:* 386-391.
100. WALLER, P. J. and R. J. THOMAS. 1975. Field studies on inhibition of *Haemonchus contortus* in sheep. *Parasitology 71:* 285-291.
101. WEBSTER, G. A. 1956. A preliminary report on the biology of *Toxocara canis* (Werner, 1782).

Can. J. Zool. 34: 725-726.
102. WHITLOCK, J. H. 1966. The environmental biology of a nematode. *In:* E. J. L. Soulsby (ed.), Biology of Parasites. Academic Press, New York.
103. WIGGLESWORTH, V. B. 1970. Insect Hormones. Freeman, San Francisco.

Use of Mathematical Models in Parasitology

ROBERT P. HIRSCH

*Division of Biology
Kansas State University
Manhattan, Kansas
U.S.A.*

INTRODUCTION

Investigations of populations and their regulation, by definition, are studies of numerical relationships. As we learn more about parasite populations, relationships of interest become more complex. As a result, it becomes exceedingly more difficult to conceptualize and communicate new hypotheses in literary language. The simplest solution to this problem is to describe these complex relationships in mathematical language. The unit of communication in the language of mathematics is the mathematical model.

As an example of the utility of modeling population concepts, let us examine a simple hypothesis. In literary language, this hypothesis is as follows:

> A population after a certain period contains, in addition to an original number of individuals, new individuals that have been born during the period of interest. The number of individuals born is a constant proportion of the original number in the population.

Although this is one of the simplest concepts in population ecology (density-independent growth), it is somewhat difficult to perceive what is taking place in such a population from the literary definition. In mathematical language, this same hypothesis becomes:

$$N_t = N_0 + \beta N_0 \qquad (1)$$

where

N_t = number of individuals after time t
N_0 = original number of individuals
β = a birth rate constant for time t

After brief inspection, anyone with minimal experience in reading mathematical language can understand how such a population will be expected to grow from time period to time period.

In addition to being a simple, yet precise mode of communicating concepts, mathematical models can serve at least two more important functions (Maynard Smith, 1968). The first of these functions has been a common motivation for those building models of populations of parasitic and disease-causing organisms: prediction of population behavior over extended periods of time. This use of models is called simulation and has a great deal of utility to parasitologists. For example, a simulation model in the hands of a scientist interested in controlling parasite populations can be used to screen several possible methods of control quickly and inexpensively. Then only the methods demonstrating the highest likelihood of success need be examined using biological models.

Gettinby (1974) used this predictive function of mathematical models to test general methods for control of populations of *Fasciola hepatica*. He investigated the effect of: (1) treatment of definitive hosts with a flukicide, and (2) control of intermediate hosts with a molluscicide. His models predicted that flukicides would be effective in relieving a particular mammalian host's parasite load, but would not have a significant effect on total parasite density within the definitive host population unless the treatment was highly efficacious. Molluscicides, on the other hand, were predicted to be effective in controlling fluke populations if they were used on a regular basis.

Another important result of employing mathematical models is that they can lead rapidly to advances in knowledge that otherwise might have to wait for accidental discovery. There are three ways in which a model might result in such a revelation. First, when a predictive model fails to fit biological data accurately, a flaw in understanding the modeled system is indicated. By dissection of such a model in light of biological data, the area of insufficient knowledge can be pinpointed. A researcher's efforts then can be directed to those problems which are in greatest need of investigation. An example of this principle in action can be found in Hairston's (1962) description of his model for populations of *Schistosoma japonicum*. This author seemed to be nearly as excited about hypotheses resulting from failure of his model to predict some types of data as he was about the success of his model to predict most data.

A second way in which models can lead to rapid increases in our knowledge of biological systems is by providing definitions of ecological principles. Examples of such models are those measuring maximum growth rates, carrying capacities, diversity, stability, rates of selection, etc. (Emlen, 1973; Maynard Smith, 1974). Although basic ecological models have not been utilized a great deal in description of parasite populations, parasitologists have used standard statistical models quite extensively. The most common of these has been that of the negative binomial distribution. This was an important component of Crofton's (1971b) model of host-parasite systems.

Use of mathematical models also can result in discovery of heretofore unknown biological principles through extrapolation or combinations of basic, well established models. In ecology, much understanding of the nature of competition and predation has resulted from manipulations of the logistic equation, a basic model of population growth (Emlen, 1973; Maynard Smith, 1974). In parasitology, Crofton (1971a) hypothesized that the basic negative binomial description of parasite distribution should appear truncated if *Gammarus pulex* survival was affected by infection with *Polymorphus minutus*. He found this truncated model an adequate description of parasite distribution where infection levels were high and concluded that host mortality had occurred. This hypothesis was later verified in the field by Pennycuick (1971).

TYPES OF MODELS

There are several ways in which a researcher can approach the problem of building a model to describe a biological system. Individual ecologists most often have adhered to one or very few techniques of modeling. Unfortunately, at times their attachment to these methods has interferred with their appreciation of the usefulness of the approach of other modelers of ecological problems. Thus far, quantitative parasite ecology is not to this point of specialization. In description of parasite populations, several parasitologists have employed what are often considered to be conflicting methods of obtaining mathematical models. If Fretwell's (1972) model of professional success is correct in its predictions, this mixing of techniques, although deleterious to our prestige, should be most efficient in advancing our understanding of parasite population dynamics.

The first decision that has to be made preparatory to mathematical description of a system is choice of the parameters of interest. There are several alternatives, the most popular of which are descriptions of number or density of organisms. Alternate methods include analysis of energy, biomass, gene frequency, or concentration of some substance. For the most part, parasitologists have been concerned with description of numbers in parasite populations. This is probably due to conceptual simplicity of working with this parameter. For example, when describing the intensity of a host's infection we most often think in terms of the number of parasites present. From the host's point of view, however, the amount of energy it had to contribute to the parasite population is probably of greater significance. These two parameters are probably not equitable, for in both intraspecific (Read, 1951) and interspecific (Holmes, 1961) competition, parasites may respond with changes in biomass (and, therefore, energy consumed) while their number is unaffected. It would be well if some parasitologists became interested in these alternative parameters in development of their models.

The major source of separation among ecological modelers is in choice of the basic method to be used to generate mathematical expressions. There are two schools of modeling which I will call descriptive and theoretical.

Descriptive modeling is concerned with finding the most accurate mathematical description of a set of biological data. Quite often this is done by the process of curve-fitting. For example, Anderson (1974) observed that feeding behavior of cyprinid fishes in Europe caused seasonal fluctuations in recruitment of larval *Caryphyllaeous laticeps* into the adult parasite populations. In order to represent this phenomenon mathematically, he employed a sine function. The components of this function were not considered in terms of their biological significance, but rather were chosen so that the model would generate values comparable to known data.

Another example of the use of descriptive models in parasitology, is reliance on the negative binomial distribution to describe over dispersion of parasites within a host population. Although hypotheses can be generated to explain why parasite distribution mimics a negative binomial (Crofton, 1971a; Whitlock, Crofton, and Georgi, 1972), no biological meaning has been assigned to the parameters of the distribution. Crofton (1971b) used the negative binomial in construction of a simulation model of a host-parasite system.

Descriptive models are the best choice of researchers interested in simulating behavior of biological systems, for they contain much information about behavior of the system in the past (Maynard Smith, 1974). The more data that is used to generate descriptive models, the greater will be their accuracy in making predictions. The major weakness of such models is the difficulty with which biological relevance is applied to their parameters. Without an interpretation of the meaning of the components of these models, we cannot increase our understanding of why systems behave as they do. This is the purpose of generating theoretical models.

In contrast to descriptive models which contain much information and are designed for simulation, theoretical models contain relatively little information and are designed to describe biological hypotheses. Theoretical models are, by intent, idealistic and oversimplified (Bartlett, 1973). This approach is necessary in order to maintain a model in a form that allows general application to a variety of biological situations. Unfortunately, this approach is also rather difficult for many biologists to accept as sound scientific method, for a theoretical model builder will begin with a set of assumptions, most of which are not universally applicable. Consequently, an astute biologist almost invariably finds exceptions to these assumptions.

To appreciate theoretical models, a biologist must first understand the philosophy behind this method of solving biological problems. We all know that biological phenomena are most often quite complex. Different species and even different individuals within a species have different ways in which to solve problems of existence. Even so, there is a degree of commonality in solutions to the same problem. This is true regardless of whether we are concerned with protein synthesis or regulation of birth rates. The theoretical model builder simply attempts a mathematical description of general quantitative phenomena. As in any generalization in biology, there will be imperfections in these models, but too much concern over these shortcomings is as injurious to advancement of knowledge as is too little concern (Skellam, 1973). When studying a model of biological systems, it is better not be become dismayed when the model fails to fit all situations, but rather to look for causes for lack of fit (Levins, 1968). At this point, validity of assumptions should be questioned for those particular circumstances that seem to be in contradiction with the model.

Development of theoretical models generally involves five steps: (1) the problem to be investigated is stated in literary language, (2) the system to be modeled is presented diagrammatically, (3) the mechanisms which are thought to control flux between compartments in the diagram are identified in literary language,

(4) these mechanisms are translated to mathematical language, and (5) the behavior of the model is compared to biological data. Examples of this process in models of parasite populations can be observed in a paper by Ratcliffe, et al. (1969).

Although the forementioned procedure will result in creation of a theoretical model, this should not be the termination of the researcher's labors. Like most good research, models should stimulate more questions than they provide answers. These questions are generally of two types. First, one should ask what would happen if the model were manipulated in some particular fashion. That is to say, what effect on the modeled population might we expect to observe if mechanisms controlling flux between compartments were altered? Secondly, because theoretical models are simplistic by nature, there will be circumstances which are not predictable. One should ask why these particular biological systems are not compatible with the model system (or, conversely, why the model was unable to predict behavior of these systems).

Both descriptive and theoretical models can be classified by yet another method. This is a distinction between deterministic and stochastic models. Most models in ecology are deterministic in nature. This means that they predict a single value for a given set of conditions. For example, Equation (1) is deterministic. For any particular N_0 and β, a single N_t is calculated. Such models fail to mirror biological systems in two ways: (1) they assume an infinite population size, and (2) they ignore random environmental fluctuations (Maynard Smith, 1974). If these shortcomings are of concern to the model builder, a stochastic model should be constructed. Rather than produce a single value for each calculated factor, stochastic models result in description of a mean value and a variance around that mean. If we were to transform the equation in the beginning of this chapter (Eq. (1)) to stochastic form, we would describe β as probabilities of 0, 1, 2, etc. offspring being born to each individual (or female) over time period t. As you might imagine, there are numerous problems with trying to predict probability functions of environmental factors. Generally, these problems lead to so much error of estimation that little advantage of stochastic models over deterministic models is realized (May, 1971).

MODELS OF PARASITE SYSTEMS

The first modelers of parasite population phenomena were interested in prediction of behavior of medically important organisms. The earliest model of this sort apparently is that of Ross (1910). This model was a not too successful attempt to predict incidence of malaria. An impressive volume of models of similar nature have appeared in the literature since that time. These epidemiological models share a common problem for biologists interested in regulation of populations of parasites. The problem is that they were constructed from the point of view of what happens to hosts rather than experiences of the parasites themselves. That is not to say that epidemiological models are not useful or uninteresting. On the contrary, some have been quite predictive and some have led to interesting hypotheses. For example, Griffiths (1971) showed by model manipulation that partial immunity in a host population can be expected to lengthen periods of disease outbreak.

In 1962, Hairston confronted the problem of constructing predictive models of *Schistosoma japonicum* from the view point of parasite population regulation. At the onset, he defined several types of data necessary to estimate parameters of this fluke's population growth. These data included indicators of definitive and intermediate host density as well as density of flukes within these hosts. Basically, Hairston's models were developed in a manner similar to that described for theoretical models. Then data were used to provide estimates of the components of the models, making them descriptive. This combination of both types of approach to model construction is common to many models of parasite systems.

Hairston's models are deterministic and probabilistic. By probabilistic I mean that the models are composed of simple frequencies of success in transferring from one compartment to another. The first model predicts the number of snails that can be expected to become infected from a given rate of egg production. That model is similar to the following:

$$R_S = (R_E)(P_H)(P_D)(P_P)(P_E) \qquad (2)$$

where
- R_S = rate of snail infection (number/day)
- R_E = rate of egg production (number/day)
- P_H = probability of an egg hatching
- P_D = probability of egg being deposited close to a snail
- P_P = probability of miracidium being able to penetrate a snail
- P_E = probability of sporocyst being able to establist in a snail

There is nothing sacred about the way in which Hairston decided to compartmentalize development to sporocysts. He legitimately could have combined some or all of the probabilities in his model into a single factor (Figure 1). By combining all probabilities,

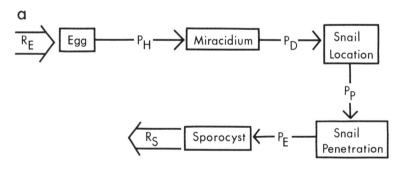

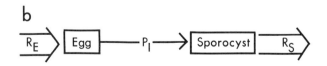

Figure 1. Comparison of (a) Hairston's (1962) original model and (b) a modified model for the snail host portion of the life cycle of Schistosoma japonicum.

this model now becomes:

$$R_S = (R_E)(P_I) \qquad (3)$$

where
- R_S = rate of snail infection (number/day)
- R_E = rate of egg production (number/day)
- P_I = probability of snail infection

The advantage of such a combination of terms is that assignment of values to model parameters becomes an easier task. In this case, one needs only to measure the rate of establishment of sporocysts for several rates of egg production. No less information is contained in this modified model. The only change I made from Hairston's original model is to combine all information in a single term. This sort of modification reduces total error of estimation by reducing the number of parameters being estimated (Bartlett, 1973). The modified model, however, has the disadvantage of containing little retrievable information. That is to say, it does not allow manipulation of components of P_I as does the original model. If Hairston had taken this more simplistic approach, he certainly would not have been stimulated to erect as many hypotheses about the nature of *S. japonicum* populations as he was when using his original model.

The second of Hairston's models was much like the first in form.

$$R_M = (R_C)(P_L)(P_M) \qquad (4)$$

where
- R_M = rate of mammalian infection (number/day)
- R_C = rate of cercariae release from snails (number/day)
- P_L = probability of a cercaria locating a mammal
- P_M = probability of maturation of a cercaria to a schistosome

Although the mammalian host portion of the *S. japonicum* life cycle has complexity comparable to the snail portion of the life cycle, Hairston chose a simpler equation to describe the former. The reason for this decision was not expressed, but we can assume that he felt there was considerable difficulty in estimation of additional mammalian parameters.

I feel there are two shortcomings of Hairston's modeling technique. First he was concerned only with *S. japonicum* populations at equilibrium. This is evident in his assumption that all parameters are stable in time. Also, another of his models states in essence that the total rate of increase of this parasite population is equal to one. Perhaps this is true of *S. japonicum* near the center of its range of distribution, but it is certainly not true in areas fringing this range. This model, therefore, cannot be used to predict behavior of parasite populations in these marginal areas.

The second shortcoming of Hairston's models is inherent to models that are composed totally of simple probabilities. This is that they ignore the possibility that rates of flux between two compartments might be influenced by the number of individuals in either compartment (i.e., density-dependence). In case of *S. japonicum*, it is likely that such regulation does play a role in population regulation (Kennedy, 1975).

Another example of models that rely entirely on probability functions are those constructed by MacDonald (1965) to predict success of male and female parasites in infecting the same host. This is obviously a problem in species which reproduce sexually and are dioecious. The simplest of MacDonald's models calculates the probability of a particular parasite finding a mate given a certain number of parasites infecting the same host.

$$P = 1 - \frac{n!}{\left(\frac{n}{2}\right)! \, 2^{\frac{2n}{2}}}$$

where
    P = probability of finding a mate
    n = number (even number or odd number - 1) of parasites infecting a host

This is simply an adaptation of basic probability concepts modified to fit the special problem of assortment of the sexes.

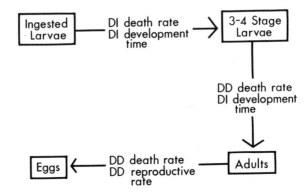

Figure 2. Summary of Haemonchus contortus *model of Ratcliffe, et al. (1969) showing density-independent (DI) and density-dependent (DD) influence on intercompartmental flux.*

Although most models of parasite populations consider only rates of flux which are not influenced by the number of individuals in either compartment (density-independent), Ratcliffe, et al. (1969) designed a model population of *Haemonchus contortus* which was subjected to both density-independent and density-dependent types of control. The basic formulation of this model involved division of the life cycle into four compartments: (1) larvae ingested by the host, (2) larvae between stages 3 and 4, (3) adults, and (4) eggs (Figure 2).

Six factors (three density-independent and three density-dependent) were considered to have direct influence on rate of flux between compartments. Between compartments one and two (larvae ingested and stage 3 - 4 larvae), both the probability of larvae development and the time required for development were considered to be independent of parasite density. Between compartments two and three (stage 3 - 4 larvae and adults), the probability of development was not a simple constant, but rather it was a function of the level of some inhibiting factor. The nature of this factor was unknown to Ratcliffe, et al. (1969), but its level was believed to be sensitive to density of parasites in compartments two and four. As the number of 3 - 4 stage larvae or the number of eggs increased, the level of inhibitor increased. As the level of inhibitor increased, the proportion of larvae which survived to maturity decreased. This defines a density-dependent death rate. The development time from compartment two to compartment three was considered to be constant. Within compartment three (adults) the second density-dependent death rate occurs. Survival of adults was considered to be sensitive to the same unknown inhibiting factor as were larvae between compartments two and three. Finally, these authors constructed a similar type of density-dependent control of the rate at which eggs were produced by adults.

This paper by Ratcliffe and his co-workers is very interesting and may be one of the best models of a parasite population that has, as of yet, appeared in the literature. Unfortunately, these authors did not report a specific mathematical description of their model, nor did they provide information as to their estimates of parameters. Undoubtedly, much effort went into arriving at this information, and it is sad that anyone interested in this parasite system will have to start without benefit of this knowledge.

Ratcliffe, et al. (1969) did not test their model. This is not unusual, nor is it a shortcoming of their study. In fact, models are probably best tested by researchers other than those who have constructed them, for it is difficult to remain totally objective in the field after so many hours have been spent affectionately with one's model. In any event, a model should be tested under conditions removed from those that aided in the model's development (Bartlett, 1973). Since this model has not been thoroughly presented, testing probably will not occur. If we are to realize the fullest returns from our modeling efforts, it is imperative that writers and reviewers of scientific papers containing mathematical models be certain that these models are in a form conducive to duplication.

Both models of Hairston (1962) and Ratcliffe, et al. (1969) were designed to simulate dynamics of populations of a particular species; and therefore, they were largely descriptive in nature. In 1971, Crofton (1971a; 1971b) approached host-parasite relationships from a more theoretical standpoint. Crofton's (1971b) presentation of his model, like the presentation of Ratcliffe, et al. (1969) was less than ideal, because the model does not appear in mathematical language. However, from his literary description of the model, we can gain some insight as to Crofton's general ideas about population dynamics. His model evidently included two functions describing limitation of a host population and three functions affecting the parasite population (Figure 3).

Crofton's host population was described as having a density-independent birth rate of two. Probably, this means that hosts had the potential of doubling their numbers every generation regardless of their density. The death rate of hosts was totally dependent on their parasite load. If the number of parasites within a host reached a predetermined lethal level, the host died. If a host never acquired this level of infection, it survived for the duration of the simulation (19 - 44 generations).

Crofton's description of functions that limited his parasite population was more vague than those for his host population. Parasite rate of reproduction and ability to infect hosts were combined into a single function he called the "Achievement Factor". Whether this "Achievement Factor" was dependent on or independent of parasite population density was not specified.

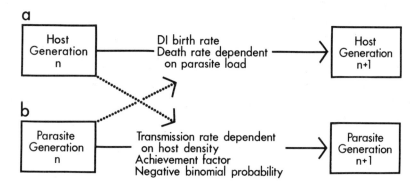

Figure 3. Summary of Crofton's (1971) model describing regulation of generalized host (a) and parasite (b) population (DI = density-independent).

Later in his paper Crofton described formulation of a parasite transmission rate as a function of host density. It is not clear if this rate was a component of the "Achievement Factor" function; or, if it was not, how transmission rate differed from ability to infect hosts. In any case, it is understood that designation of transmission rate as a function of host density and host survival as a function of parasite load caused both populations to be regulated in a density-dependent fashion. If parasite number increased, host mortality increased, transmission rate decreased, and parasite number decreased. By the same token, if host number increased, parasite transmission increased, host mortality increased, and host number decreased. With such strong interdependence of these populations, it is not surprising that Crofton's model produced stable densities of hosts and parasites.

The final function influencing the parasite population in Crofton's model was a negative binomial description of the distribution of parasites among hosts. Exactly how this probability function was used in the model was not specified, but it was most likely used to calculate the number of parasites that established in a host after exposure to infection. Although the negative binomial has been well established as an adequate descriptor of distribution of parasites within host populations, I believe it is risky to use this probability function in a predictive model as just described.

The danger in reliance on the negative binomial probability function results from the fact that this distribution is generated at the suprapopulation level, but Crofton's model is applied at the infrapopulation level. Since we do not know what, if anything, is happening at the level of the individual host to create or participate in the creation of a negative binomial in the suprapopulation, this probability function should not be used to predict events in infrapopulations. For example, it is likely that hosts that harbor the heaviest infections are those hosts who are most resistant to that level of infection becoming lethal. In other words, the "tail" of the negative binomial might reflect the percentages of the host population that were exposed to heavy parasite loads and survived. If these percentages are used to predict the probability that a host will acquire a certain parasite population, this probability will be underestimated. Further error results when a lethal level is applied to these resistant hosts. Such confounding of lethal levels and the negative binomial would seriously affect the model's ability to mirror reality.

Anderson (1974a; 1974b) developed somewhat general models to describe various types of host-parasite interactions. The first of these was a stochastic model constructed mostly in a descriptive manner. However, the model is more generally applicable than most descriptive models for this one has a theoretical framework. Basically, the model is a variation of a standard expression of exponential growth (Wilson and Bossert, 1971) described in Equation (6).

$$\frac{dN_t}{dt} = bN_t - dN_t \qquad (6)$$

where

N_t = number of organisms in a population at a particular time
b = birth rate per individual over time t
d = death rate per individual over time t

Anderson's model (Figure 4) is similar to the following:

$$\frac{dN_t}{dt} = i - dN_t \qquad (7)$$

where

N_t = number of parasites in an infrapopulation at a particular time
i = immigration rate per host over time t
d = death rate per parasite over time t

There are two differences between Equations (6) and (7). The first one is obvious. Instead of a birth rate as in Equation (6), Equation (7) contains an immigration rate. This is not a very important difference, for birth can be considered a type of immigration. More importantly, Anderson's immigration rate describes events occurring at the infrapopulation level whereas

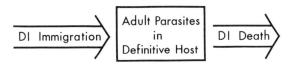

Figure 4. Summary of Anderson's (1974) model describing regulation of an infrapopulation of Caryophyllaeus laticeps *in the definitive host (DI = density-independent).*

the traditional birth rate occurs at the individual level. Anderson recognized that immigration is probably a function of the birth rate of individuals in the suprapopulation at some previous time, but his model was limited to infrapopulations. Since there was no way to predict what was happening in the suprapopulation from parasite density in the infrapopulation, he chose to describe immigration as a random variable.

To define changes in immigration and death rates occurring over time, Anderson changed from a theoretical to a descriptive approach. As previously stated in this chapter, in order to construct a descriptive model, it is necessary to have a data base. Anderson chose data he had collected on *Caryophyllaeus laticeps*, a cestode of fish (Anderson, 1971). It had been his observation that immigration of these parasites into a host was a seasonal phenomenon, dependent on host feeding behavior. Therefore, he chose a cyclic function to describe immigration rate.

$$i = a + b \left[\sin \frac{2 \pi (t - \tau)}{12} + 1 \right] \qquad (8)$$

where
t = time
τ = a phase angle
a and b = constants

Equation (8) was chosen because it provided a good fit to observed data; not because it expressed some particular biological hypothesis. Although Anderson did not attempt to assign biological meaning to his parameters (e.g., a and b), it would be interesting to do so. This step would be necessary, in fact, if the model were to be used in prediction of unobserved events in an infrapopulation.

The same process of curve fitting led Anderson to develop an expression for death rate. He had previously (Anderson, 1971) observed that death of *C. laticeps* was correlated with water temperature. The equation which best fits his data was as follows:

$$d = ae^{bT} + c \qquad (9)$$

where

T = water temperature
a, b and c = constants

Anderson further observed that water temperature was related to season, and therefore, developed the following expression of death as a function of time:

$$d = ae^{\left[b\left(x + y\sin\frac{2t}{12} + z\cos\frac{2t}{12}\right)\right]} + c \qquad (10)$$

where

t = time
a, b, c, x, y and z = constants

To this point, Anderson's model and its components were in deterministic form, yet it was Anderson's (1974b) objective to develop a stochastic model. To make the transition from deterministic to stochastic, parameters expressed as absolute values must be expressed as means with some distribution around these means (reported as variances). Anderson converted his deterministic expression of immigration rate to stochastic form by using field data to estimate variances for the constants in Equation (8). Stochastic relationships were then used in Poisson description of the number of parasites to be expected to infect a given host at some time t.

$$\overline{N}_t = \overline{a} \int_{t-s}^{t} g(u)\,du + \overline{b} \int_{t-s}^{t} s(u)g(u)\,de \qquad (11)$$

and

$$\sigma^2_{N_t} = \sigma^2_a \left[\int_{t-s}^{t} g(u)\,du\right]^2 + \sigma^2_b \left[\int_{t-s}^{t} s(u)g(u)\,du\right]^2 \qquad (12)$$

where
$\overline{N2}_t$ = mean number of parasites at time t
$\sigma^2_{N_t}$ = variance of N_t
$s(u) = \sin \dfrac{2\pi(t - \tau)}{12} + 1$
$g(u) = e^{(-\int_u^t \mu(v)\,dv)}$
$\overline{a}$ = mean of a from equation 7
$\overline{b}$ = mean of b from equation 7
σ^2_a = variance of a from equation 7
σ^2_b = variance of b from equation 7
s = maximum life span of adult parasite

Armed with these models of *C. laticeps* infrapopulations, Anderson (1974b) tested two hypotheses. The first of these was that both the immigration rate and death rate must be expressed as a function of time in order to more closely imitate behavior of actual infrapopulations. This test involved the original model (Eq. (11)) and two variations of this model. One variation expressed death rate as a function of time and immigration rate as a constant. The other expressed death rate as a constant and immigration rate as a function of time. All three expressions of infrapopulation density had a cyclic nature approximating values observed in the field. Although no statistical test was applied to determine how well these models fit observed data, it appears from my inspection that the worst fit was the model which expressed only death as a function of time. This indicates the inadequacy of seasonal death, and therefore, the importance of seasonal feeding behavior of the host in determining the size of infrapopulations. It is difficult, however, to choose which of the remaining two models was the best fit of observed data without a statistical test. Anderson did not speculate as to which model was best, but it appears to me that the model which expressed immigration as a function of time and death as a constant better predicted field data than did the original model. In any case, seasonal death does not seem to be an important factor in determining the size of infrapopulations of *C. laticeps*.

Another hypothesis examined by Anderson was that there was little influence of random fluctuation in temperature on stability of *C. laticeps* infrapopulations. To test this hypothesis, Anderson used a random number generator to choose values of x, y, and z in Equation (10). This treatment did not significantly effect the mean number of parasites expected per host, supporting Anderson's hypothesis.

To achieve a general description of a complete parasite life cycle, Anderson (1974b) considered the direct life cycle from an unusual point of view. Traditionally, a parasite population is compartmentalized according to life cycle states (egg, larva, adult, etc.) and models describe factors controlling flux between compartments. In contrast to this approach, Anderson chose to compartmentalize the host population according to parasite load. That is to say, he treated hosts with 0, 1, 2, 3 .... n parasites infecting them as separate "populations". These "populations" could increase by the death of one parasite within a host belonging to the "population" of hosts with the next higher increment of infection or by immigration of one parasite into a host belonging to the "population" of hosts with the next lower increment of infection. For example, consider the "population" of hosts defined by the fact that they are infected by 10 parasites. The number of hosts in this "population" could increase by immigration of one parasite into a host containing nine parasites or by death of one parasite within a host containing 11 parasites. Either event would increase the number of hosts in the 10-parasite "population" by one. Simultaneously, the number of hosts in the 9-parasite "population" or 11-parasite "population" would decrease by one.

As well as changes in parasite load affecting number of hosts in various "populations", hosts were born into the 0-parasite "population" and died out of all "populations". The total number of individuals in the parasite suprapopulation was determined by summing the number of hosts in each infection "population" multiplied by the number of parasites infecting each host in that "population".

$$P_T = \sum_{P=0}^{n} PH_P \qquad (13)$$

where
P_T = total number of individuals in parasite suprapopulations
P = number of parasites in each infrapopulation
n = maximum number of parasites possible in an infrapopulation
H_P = number of hosts in P-parasite "population"

Anderson first assumed that all rates of change were constant and, therefore, density-independent. With this assumption he found his model to be very unstable. To alleviate this problem, Anderson chose to make host mortality dependent on host density (density-dependent mortality) and on the number of parasites infecting each host. He assumed an exponential function.

$$d_h = ae^{(bP + cH_T)} \qquad (14)$$

where
d_h = death rate of hosts per individual
P = number of parasites in infrapopulation
H_T = total number of hosts in all "populations" combined
a, b, and c = constants

Incorporation of this relationship (Eq. 14) into Anderson's model produced stability in the form of convergent oscillations in both host and parasite populations. He found that factors which contributed most to determination of equilibrium values were the maximum parasite load, rate of immigration of eggs into adult parasite populations and rate of egg production by adult parasites. If the latter two factors became too large, parasite populations risked extinction. By expansion of his model to describe an indirect life cycle with one intermediate host, Anderson demonstrated how higher parasite birth rates and immigration rates could be tolerated without permanent elimination of the parasite population.

Anderson's manipulations of his models exemplify the process of hypothesis testing using mathematical descriptions of natural populations. The researcher asks, "What would happen to my modeled population if I performed the following manipulation?" The appropriate adjustments are made to the model and resultant behavior of the theoretical population is observed. The next step should be comparison of the modeled population's behavior with that of natural populations. If the model and nature agree, we can allow ourselves greater confidence in the model's ability to mirror reality. If the model and nature are in conflict, the question should be asked, "What is it about nature that we understand so poorly that we were unable to construct a reliable model?" The answer to this question is generally found by careful inspection of the model and its assumptions until the source of error is discovered. As previously stated in this chapter, it is this process of "trouble-shooting" which most often leads to interesting discoveries. In the case of Anderson's model of a direct life cycle-type of parasite population, it should be quite enlightening to explore the reasons this model failed to produce a negative binomial distribution except under the most limited conditions. Perhaps by studying this problem we can gain much needed knowledge regarding the biological factors producing this distribution with such consistency in parasite suprapopulations.

The final model chosen for discussion in this section is that of Gettinby (1974). The approach of this researcher offers an example of mathematical model developed to answer some very well defined questions about a single species of parasite. Specifically, Gettinby was interested in comparing two methods of controlling *Fasciola hepatica*: (1) use of flukicide to reduce adult flukes within their mammalian hosts and (2) use of a molluscicide to reduce intermediate hosts in the environment.

Often projects of this sort involve use of descriptive models so that maximum predictability can be achieved. In order to develop a descriptive model, however, a large quantity of data is necessary from which to generate a mathematical representation of population dynamics. Evidently sufficient data of the type required for model building does not exist for *F. hepatica*; thus, Gettinby chose an approach resembling theoretical modeling.

Gettinby's model is concerned only with development of *F. hepatica* from miracidia to sporocysts (Figure 5). We must assume, therefore, that it was the author's contention (unstated) that the remainder of the fluke life cycle can be described by some linear probability function. Gettinby divided the modeled life cycle segment into two, temporally distinct, probabilistic events. The first of these is the probability of a miracidium encountering a snail. The second is the probability that a miracidium, which has located an intermediate host, will successfully penetrate into (and, we might assume, establish within) that host.

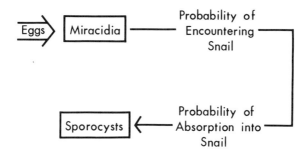

Figure 5. Summary of Gettinby's (1974) model of Fasciola hepatica *populations associated with the intermediate host.*

The first portion of Gettinby's model describes a situation in which snails are randomly distributed in some area of the environment. A cluster of miracidia (from eggs in a single fecal deposition) is located at some point within this same environment. Miracidia emigrate from this cluster in a symmetrical manner, having equal probability of traveling in any direction within 360°. The distance these miracidia travel from the original cluster is predicted from an exponential distribution about some mean distance. Since miracidia are able to chemotactically locate snails if they are within 15 cm, encounter is assumed to take place if, after random emigration, a miracidium and snail are at least this close to each other.

Mathematically, this portion of Gettinby's model was built on a joint probability density function describing final positions of miracidia relative to their original cluster measured on polar co-ordinates.

$$p(r,\Theta)drd\Theta = (\lambda e^{-\lambda r} dr)(\frac{1}{2\pi} d\Theta) \qquad (15)$$

where

$p(r,\Theta)$ = probability that miracidium will have travelled r cm in direction $\Theta°$
r = distance traveled
Θ = direction traveled
λ = inverse of mean distance miracidia can travel

The portion of Gettinby's model describing absorption of miracidia into snails they have encountered was based on the sum of two probability functions. The probability that a certain number of miracidia are absorbed over a period of time was taken to be the sum of the probability that the last miracidium is absorbed in the last increment of time and the probability that all miracidia are absorbed before the last increment of time. This sort of reasoning led to a binomial probability distribution.

$$P_r(t) = \frac{n!}{(n-r)!r!} e^{-\beta(n-r)t} \left(1-e^{-\beta t}\right)^r \qquad (16)$$

where
    $p_r(t)$ = probability that r miracidia are absorbed at time t
r = number of miracidia absorbed
n = number of miracidia encountering snail
β = absorption rate
t = time (taken as life span of miracidia)

The first step to prediction, once a model is formulated, is estimation of values of model components. In the case of Gettinby's model, components requiring assignment of values are: (1) $1/\lambda$ or the mean distance miracidia can travel, (2) β or the rate at which miracidia are absorbed into snails, (3) the density of snails in the environment, and (4) the number of miracidia which occur in original clusters (fecal deposits). The author used previously acquired data to estimate values for the first 2 components ($1/\lambda$ and β). These data, however, were quite limited so Gettinby tested the sensitivity of his model to variation in estimates of $1/\lambda$ and β. This test was accomplished by random choice of values to be assigned to each component. The range of values available for selection in such a sensitivity analysis should encompass all possible real quantities of the components. Through this procedure, Gettinby discovered that it made little difference to his model's predictions whether these two components were fixed at some measured value or varied within a range of values.

    Evidently, Gettinby did not have data available for estimation of snail miracidium density. In the circumstance that estimates cannot be made for model components, it is necessary to utilize a wide range of values for these components in prediction. In this case, Gettinby chose 0.5 to 20 snails per m^2 for snail density and 50 to 10,000 miracidia per cluster for miracidium density. Since data did not contribute to selection of these extremes for model components, we cannot weight too heavily behavior of the model over a short range of values, especially near either extreme.

    Initial conditions for simulation of Gettinby's model populations (position of snails and miracidium clusters) were chosen at random. Both encounters and absorptions were found to be linearly related to snail density. Prediction of encounters and absorptions over several miracidium cluster densities, however, did not result in a linear relationship. Instead, Gettinby found that encounters and absorption increased in frequency as cluster density increased from 50 to about 3000 miracidia per cluster. There was little change in frequency of encounters and absorptions as cluster density increased beyond 3000 miracidia. This observation led Gettinby to conclude that control measures which reduce cluster density (i.e., flukicides) would not be able to affect incidence of infection unless they were capable of maintaining cluster density below 3000 miracidia. Gettinby felt that a flukicide would not be very successful in reducing cluster density to this degree, so he recommended use of a molluscicide to reduce incidence of infection with *F. hepatica*. According to predictions made by this model, reducing snail density should cause a corresponding decrease in snail (and, therefore, mammal) infection rate for all snail densities tested.

DEVELOPMENT OF A MODEL

As an example of the processes leading to development of a mathematical model, I present in this section a model I have developed for definition of an ecological principle which, through extrapolation, suggests an interesting hypothesis. The formulation of this model was stimulated by my curiosity about the type of regulation to which larval *Taenia pisiformis* populations are exposed within an intermediate host. I divided development of this larval tapeworm into three stages: onchospheres, liver stage larvae, and infective cysticerci (Figure 6). My plan was to administer various dosages of onchospheres to rabbits and measure the success of generation of subsequent stages. Specifically, I was interested in determining if density-dependent regulation occurred during development from onchospheres to liver stage larvae and/or from liver stage larvae to infective cysticerci. It immediately became obvious that the latter of these two intervals would provide difficulty in measurement. It is an easy matter to administer onchospheres to rabbits and

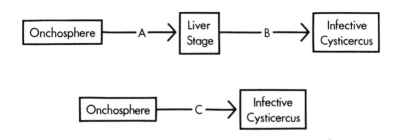

Figure 6. Life stages of larval Taenia pisiformis. *Regulation over A and B are in question, but regulation over only A and C can be directly measured.*

observe liver stage larvae and cysticerci yields, but it is not possible (to my knowledge) to administer liver stage larvae to rabbits and observe patterns of cysticerci yield. It is necessary, therefore, to determine the pattern of regulation between liver stage larvae and infective cysticerci by some indirect method.

I felt that the solution to this problem was to develop a mathematical model which would predict yields of cysticerci expected after rabbits had received various dosages of onchospheres. This model would have to demonstrate any differences that would occur if: (1) the number of liver stage larvae resulting from various dosages of onchospheres is density-dependent and the number of infective cysticerci resulting from various levels of liver stage larvae is density-independent, (2) the number of liver stage larvae resulting from various dosages of onchospheres is density-independent and the number of infective cysticerci resulting from various levels of liver stage larvae is density-dependent, (3) no density-dependent regulation occurs during development from onchospheres to cysticerci, or (4) density-dependent regulation occurs during both stages of development from onchospheres to cysticerci. I started with a basic linear model of population growth. Since no birth is assumed to occur from onchospheres to cysticerci[1], I concerned myself only with death rates. The basic model is:

$$N_t = N_0 - dN_0 \qquad (17)$$

where
  N_0 = number of individuals at stage 0
  N_t = number of individuals at stage t
  d = death rate/individual from 0 to t

If the death rate is independent of initial density (N_0), then the death rate is equal to some constant value. This assumes that what ever causes death is also constant (e.g., environmental conditions). This is probably not exactly true, but deviation from constancy will not affect basic deductions to follow. Mathematically, a density-independent death rate can be simply represented as:

$$d = C_0 \qquad (18)$$

where
  C_0 = some constant

Graphically, a density-independent death rate plotted against density (N) appears as a line with a slope equal to 0 (Figure 7). That is to say, as population density increases, death remains unchanged.

If death rate is density-dependent, it is equal to some function of the initial population density. Mathematically, this function can be represented by an equation describing a linear relationship.

$$d = C_1 + C_2 N_0 \qquad (19)$$

where
  C_1 and C_2 = constants

---

[1]Recent experiments on *T. pisiformis* have suggested that asexual reproduction of cysticerci does occur *in vitro* (Heath, 1973). It is not certain whether similar reproduction occurs *in vivo*. In any case, larval *T. pisiformis* populations behave as if no reproduction occurs.

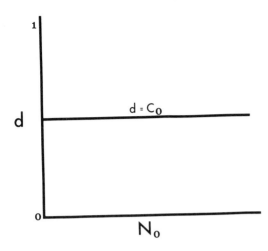

Figure 7. A density-independent death rate.

Graphically, a density-dependent death rate as a function of density appears as a line with a slope (C_2) not equal to zero (Figure 8). In most cases the slope is positive meaning that, as population density increases, the rate of death (per individual) increases.

To obtain an expression of the density of individuals expected in generation t following density-independent death, Equation (18) is substituted into Equation (17).

$$N_t = N_0 - C_0 N_0 \qquad (20)$$

Equation (20) is not different from the original equation (Eq. (17)) in form, but if we substitute Equation (19) into Equation (17) to obtain an expression of N_t following density-dependent death, a second order polynomial results.

$$N_t = N_0 - C_1 N_0 - C_2 N_0^2 \qquad (21)$$

These manipulations cause us to hypothesize that, if the number of cysticerci found as a function of the number of onchospheres administered to rabbits is a linear relationship, no density-dependent regulation has occurred; whereas, if this function is a second order polynomial, there has been a density-dependent

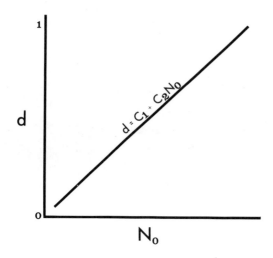

Figure 8. A density-dependent death rate.

regulating factor operating at some point between these stages of development. This result is neither original nor is it a complete solution to the problem at hand. What if two or more density-dependent events occur during development from onchospheres to infective cycticerci?

Let us assume for the moment that two separate events during the life cycle of *T. pisiformis* regulate numbers in a density-dependent manner. Let us further assume that one density-dependent event occurs during development from onchospheres to liver stage larvae and the other from liver stage larvae to infective cysticerci. Then, using Equation (21), the number of liver stage larvae developing from a given dosage of onchospheres can be described by:

$$N_L = N_0 - C_1 N_0 - C_2 N_0^2 \qquad (22)$$

where
N_0 = number of onchospheres administered
N_L = number of liver stage larvae
C_1 and C_2 = constants

Again using Equation (21), the number of infective cysticerci developing from a certain number of liver stage larvae can be described by

$$N_C = N_L - C_3 N_L - C_4 N_L^2 \qquad (23)$$

where
   N_C = number of infective cysticerci
   C_3 and C_4 = constants

An expression of the number of infective cysticerci which is expected to develop from a particular dosage of onchospheres can be obtained by substituting Equation (22) into Equation (23). This manipulation results in the following expression.

$$N_C = C_5 N_0 - C_6 N_0^2 + C_7 N_0^3 - C_8 N_0^4 \qquad (24)$$

where
   $C_5 = 1 - C_1 - C_3(1-C_1)$
   $C_6 = C_2(1-C_3) + C_4(1 - C_1)^2$
   $C_7 = 2 C_2 C_4 (1 - C_1)$
   $C_8 = C_2^2 C_4$

Equation (24) states, if two, temporally distinct, density-dependent events occur during development from onchospheres to infective cysticerci, cysticerci will be a fourth order polynomial function of onchospheres. It can be demonstrated further that this basic relationship is unaltered by any number of density-independent events occurring during this same period. Continuing this process, polynomials of increasing order are predicted for additional density-dependent events. The general relationship is:

$$D = \log_2 P \qquad (25)$$

where
   D = number of density-dependent events
   P = order of polynomial

Now that this mathematical definition has been developed, it is an easy matter to fit population data to various polynomials. The order of the polynomial which best fits these data determines how many, if any, density-dependent events have occurred between two points in development. A detailed account of this process applied to data obtained for larval *T. pisiformis* is being prepared for publication elsewhere. In essence, it seems as if *T. pisiformis* experiences one density-dependent episode in its development from onchospheres to infective cysticerci.

Although this model was developed as a tool to be used in assessment of population data, its usefulness need not end with this function. As stated in a previous section of this chapter, model manipulation and hypothesis testing can lead to interesting observations. Often, these observations are of greater significance in increasing our understanding of population regulation than is the solution to the problem which motivated the model's development. I believe this to be true of the model presently under consideration.

For purposes of hypothesizing, we will consider more general application of the principles previously elucidated. Let us assume we are studying a population of life stage organisms which experience density-dependent death rates during two development periods (Figure 9); one between eggs and larvae (A) and the other between larvae and adults (B). Let us further assume that the rate at which eggs are produced by adults is independent of population density. From Equation (19), we know that the death rate from eggs to larvae can be expressed as:

$$d_A = C_1 + C_2 N_E \qquad (26)$$

where

d_A = death rate/individual from eggs to larvae
N_E = number of eggs
C_1 and C_2 = constants

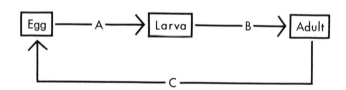

Figure 9. Life stages of a theoretical population with flux between stages controlled by density-dependent death (A + B) and density-independent birth (C).

Similarly, the death rate occurring during development from larvae to adults can be described by

$$d_B = C_3 + C_4 N_L \tag{27}$$

where
d_A = death rate/individual from larvae to adults
N_L = number of larvae
C_3 and C_4 = constants

From Equation (17), we also know that the number of larvae in the population should be a function of the previous number of eggs and the rate of death development from eggs to larvae. Specifically, this function is:

$$N_L = N_E - d_A N_E \tag{28}$$

By substituting Equation (26) for d_A in Equation (28) and then substituting the resultant expression for N_L in Equation (27), a description of d_B in terms of N_E is obtained.

$$d_B = C_3 + (C_4 - C_1 C_4) N_E - C_2 C_4 N_E^2 \tag{29}$$

These density-dependent death rates now can be combined to describe the total rate of mortality from eggs to adults. Total death can be expressed as:

$$d_T + 1 - (1 - d_A)(1 - d_B) \tag{30}$$

where
d_T = total death rate/individual from egg to adult

Substituting Equation (26) for d_A and Equation (29) for d_B in Equation (30) yields the following expression of d_T in terms of N_E.

$$d_T = C_5 + C_6 N_E - C_7 N_E^2 + C_8 N_E^3 \tag{31}$$

where
$C_5 = C_1 + C_3(1-C_1)$
$C_6 = C_2(1-C_3) + C_4(1-C_1)^2$

$$C_7 = 2C_2C_4(1-C_1)$$
$$C_8 = C_2^2 C_4$$

It is comforting to note at this point that the third order polynomial we have obtained as an expression of the death rate resulting from two density-dependent events acting in sequence is one minus the first derivative of the expression of number of organisms under similar conditions (Eq. (24)). As we learned in Freshman Calculus, the rate of change of some function can be found by determination of its first derivative.

As previously stated, Figure 8 is the graphical representation of the commonly accepted mathematical expression of density-dependent death. On many occasions, researchers who have tried to fit death rate data to similar straight lines have found a poor fit (Emlen, 1973). Instead, they have found their data to best fit an asymptotic or even sigmoid curve. It is interesting, in this light, to consider Figure 10, a graphical representation of Equation (31). Depending on the value of the constants in Equation (31), this curve could fit these previously observed data quite well. Perhaps, unknowingly, these researchers were studying populations which are controlled by multiple density-dependent events.

As previously assumed, our theoretical population has a rate of egg production which is independent of population density. This means that the birth rate can be described simply by a constant (see Eq. (18)).

$$b - C_9 \tag{32}$$

where

C_9 = constant

A population is said to be at equilibrium when its birth rate is equal to its death rate. This means that the population is neither increasing nor decreasing when it contains the number of organisms referred to as the "equilibrium density." The population will remain at this level as long as conditions remain stable. Equilibrium densities can be determined mathematically by finding solutions to Equation (34).

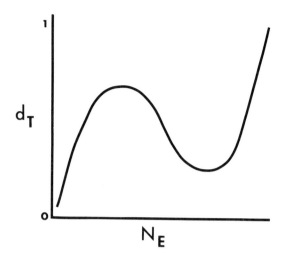

Figure 10. *Graphical representation of death rate as a third order polynomial function of the number of organisms in a population.*

$$b = d \qquad (33)$$

or

$$c_9 = c_5 + c_6 N_E - c_7 N_E^2 + c_8 N_E^3 \qquad (34)$$

There are three equilibria which are solutions to Equation (34). Two of these equilibria are stable equilibria and one is an unstable equilibrium. Minor deviations from stable equilibrium densities stimulate changes in birth and/or death rates which cause the population density to continue to change in the direction of the original deviation. This is probably best demonstrated graphically.

Figure 11 shows a plot of both birth and death rate as functions of population level (N_E was chosen for convenience). An equilibrium is indicated wherever these two curves intersect. At values of N_E equal to A or C, the population is at a stable equilibrium. If N_E increases from either of these values, it can be seen that the death rate will increase to become greater than the birth rate causing the population to return

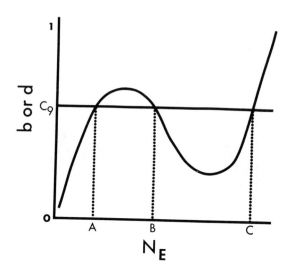

Figure 11. Graphical representation of equilibria which could result from two sequential density-dependent death rates and one density-independent birth rate. Two of the equilibria (A + C) are stable while the third (B) is unstable.

to the equilibrium density. The converse is true for decreases from these equilibria. If N_E is equal to B, on the other hand, the population is at an unstable equilibrium. It can be seen readily that an increase or decrease in N_E from B will cause changes in the death rate which will stimulate further deviation from B. Through time, a population such as this might fluctuate in a manner similar to that demonstrated in Figure 12.

In other words, there are two levels (A and C) around which such a population will fluctuate (due to fluctuating environmental conditions). Most likely the usual population size will correspond to values near N_E equal to A, but occasionally fluctuations in numbers will lead to a population with N_E greater than B. When this happens, the population will continue to increase and then fluctuate around N_e equal to C until N_E once again falls below B. One would predict, therefore, that such a population would occasionally reach "epidemic" levels.

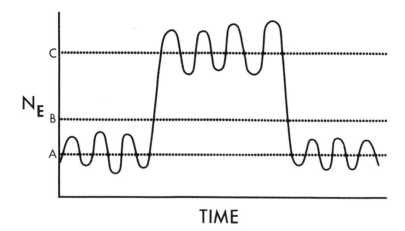

Figure 12. *Dynamics possible for a population with one unstable (B) and two stable (A + C) equilibria.*

Although this hypothesis has been developed from the infrapopulation point of view, it is possible that it could, if verified by observation, contribute to our knowledge of both infrapopulation and suprapopulation phenomena. For example, at the infrapopulation level, it seems as if parasite levels are usually regulated in some density-dependent manner (Kennedy, 1975). Occasionally, however, parasites in infrapopulations can become quite numerous. Although it is often times unclear what factors are regulating parasite numbers, it seems as if exceptionally heavy infections are associated with extremely heavy exposure to infective stages. Such a nonlinear relationship between exposure and infection is suggestive of a multi-equilibria system not unlike the one hypothesized. This system also would be expected to result in a distribution of parasites in the suprapopulation in which most hosts contain few parasites while few hosts contain most of the parasites. This describes, in part, a negative binomial distribution.

LITERATURE CITED

ANDERSON, R. M. 1971. A quantitative ecological study of the helminth parasites of the bream *Abramis brama* (L). Ph. D. thesis, London University

ANDERSON, R. M. 1974a. Population dynamics of the cestode *Caryophyllaeus laticeps* (Pallas, 1781) in the bream (*Abramis brama* L.). J. Ani. Ecol. 43: 305-321.

ANDERSON, R. M. 1974b. Mathematical models of host-helminth parasite interactions. *In:* M. B. Usher and M. H. Williamson (eds.), Ecological Stability. Chapman and Hall, London.

BARTLETT, M. S. 1973. Equations and models of population change. *In:* M. S. Bartlett and R. W. Hirons (eds.), The Mathematical Theory of the Dynamics of Biological Populations. Academic Press, New York.

CROFTON, H. D. 1971a. A quantitative approach to parasitism. *Parasitology* 62: 179-193.

CROFTON, H. D. 1971b. A model of host-parasite relationships. *Parasitology* 63: 343-364.

EMLEN, J. M. 1973. Ecology: An Evolutionary Approach. Addison-Wesley, Reading and Menlo Park.

FRETWELL, S. D. 1972. Populations in a Seasonal Environment. Princeton University Press, Princeton.

GETTINBY, G. 1974. Assessment of the effectiveness of control techniques for liver fluke infection. *In:* M. B. Usher and M. H. Williamson (eds.), Ecological Stability. Chapman and Hall, London.

GRIFFITHS, D. A. 1971. Epidemic models. Ph. D. thesis, University of Oxford.

HAIRSTON, N. G. 1962. Population ecology and epidemiological problems. *In:* G. E. W. Wolstenholme and M. O'Connor (eds.), Bilharziasis. Little, Brown and Company, Boston.

HEATH, D. D. 1973. An improved technique for *in vitro* culture of taeniid larvae. *Internat. J. Parasitol.* 3: 481-484.

HOLMES, J. C. 1961. Effects of concurrent infections on *Hymenolepis diminuta* (Cestoda) and *Moniliformis dubius* (Ancanthocephala). I. General effect and comparison with crowding. *J. Parasitol.* 47: 209-216.
KENNEDY, C. R. 1975. Ecological Animal Parasitology. John Wiley and Sons, New York.
LEVINS, R. 1968. Evolution in Changing Environments. Princeton University Press, Princeton.
MACDONALD, G. 1965. The dynamics of helminth infections with special reference to schistosomes. *Trans. Royal Soc. Trop. Med. Hyg.* 59: 489-506.
MAY, R. M. 1971. Stability in model ecosystems. *Proc. Ecol. Soc. Aust.* 6: 18-56.
MAYNARD SMITH, J. 1968. Mathematical Ideas in Biology. Cambridge University Press, Cambridge.
MAYNARD SMITH, J. 1974. Models in Ecology. Cambridge University Press, Cambridge.
PENNYCUICK, L. 1971. Frequency distributions of parasites in a population of three-spined sticklebacks, *Gasterosteus aculeatus* L., with particular reference to the negative binomial distribution. *Parasitology* 63: 389-406.
RATCLIFFE, L. H., H. M. TAYLOR, J. H. WHITLOCK, and W. R. LYNN. 1969. Systems analysis of a host-parasite interaction. *Parasitology* 59: 649-661.
READ, C. P. 1951. The "crowding effect" in tapeworm infections. *J. Parasitol.* 37: 174-178.
ROSS, R. 1910. The Prevention of Malaria. John Murray, London.
SKELLAM, J. G. 1973. The formulation and interpretation of mathematical models of diffusionary processes in population biology. *In:* M. S. Bartlett and R. W. Hiorns (eds.), The Mathematical Theory of the Dynamics of Biological Populations. Academic Press, New York.
WHITLOCK, J. H., H. D. CROFTON, and J. R. GEORGI. 1972. Characteristics of parasite populations in endemic trichostrongylidosis. *Parasitology* 64: 513-527.
WILSON, E. O. and W. H. BOSSERT. 1971. A Primer of Population Biology. Sinauer Associates, Stamford, Conn.

Populations in Perspective: Community Organization and Regulation of Parasite Populations

JOHN C. HOLMES, RUSSELL P. HOBBS and TAK SENG LEONG

Department of Zoology
University of Alberta
Edmonton, Alberta
Canada

INTRODUCTION

The term "population", as applied to parasites, is ambiguous. One may legitimately wish to denote at least two levels of groupings: all the individuals of a single species in a single host individual, or all the individuals of all stages of a single species in an ecosystem, whether they are free-living or in their intermediate or definitive hosts. In this paper, we follow Esch, Gibbons, and Bourque (1975), who used the terms "infrapopulation" and "suprapopulation" for these two groupings.

The regulation of parasite populations has been investigated at both levels. Most investigators have concentrated on regulation at the infrapopulation level, since most of the classic mechanisms, such as immunity or competition, work at that level, and since regulation at the infrapopulation level can lead to regulation at the suprapopulation level. In addition, it is at the infrapopulation level that high populations can produce disease.

However, it is the suprapopulation level at which regulation is the most significant. Parasitologists working at that level generally recognize that different mechanisms may be operating on different stages in the life cycle of the parasite. There is less recognition that different mechanisms may be operating on the infrapopulations of the same stage, but harbored by different species of hosts. It is this aspect of regulation that we will explore here.

Regulation (used as in Kennedy's paper in this volume) of any population requires the interplay of two conditions:
1) a reproductive potential adequate to increase the population size, and
2) some form of negative feedback control adequate to prevent that increase in population.

Bradley (1972), in an excellent paper on the theory of regulation of parasite populations, pointed out that the first condition is fundamental, but that an examination of the second condition is more profitable for an understanding of regulation. He went on (1972; 1974) to outline three types of ways in which the upper limits of parasite populations may be determined (not necessarily regulated):
1) by transmission (Type I), with no negative feedback mechanisms operative, therefore no regulation in the sense used here,
2) by regulation at the level of the host population (Type II), through a "sterile" immune response, which eliminates the parasites in individual hosts and prevents re-infection (as in classical models of epidemics, Bailey, 1957), or through overdispersion of the parasites within the host population, with subsequent death of the most heavily infected individuals (Crofton, 1971a; 1971b; Pennycuick, 1971a; 1971b), and
3) by regulation at the level of the host individual (or the parasite's infrapopulation) (Type III), through some form of partial immunity.

Bradley has suggested that transmission-determined infection (Type I) is precarious and unstable, but that it may characterize parasites at the edge of their range or parasites of hosts with short life spans, in which "annual replacement of the host stock ensures that parasite populations do not build up over very long periods of time" (1974).

Bradley considers that Type II regulation is also precarious in the real world, since both sterile-immune or aggregation-mortality mechanisms are highly dependent on the size and spatial structure of the host population. He therefore regards Type III as "the ultimate in regulation: highly efficient transmission combined with 'premunition' or some similar process of parasite population regulation *by each host*" (1974; emphasis ours). His examples, the immune inhibition of reproduction of *Trypanosoma lewisi* (Taliaferro, 1924; D'Alesandro, 1962), *Leishmania donovani* (Bradley, 1971), or trypanosomes of the *brucei* subgroup (Gray, 1965), and the "concomitant" immunity to schistosomes (Smithers, Terry, and Hockley, 1969), all involve immune responses.

Immune responses are probably the most important class of mechanisms in Type III regulation, but they are not the only ones possible. Clearly, regulation *at the level of* the host individual cannot be equated with regulation *by* the host individual. Nonimmunological mechanisms, similar to those which regulate populations of free-living organisms, may also apply to parasites.

The most obvious and well-known of these mechanisms in parasites is intraspecific exploitative competition, or the "crowding effect," best known in tapeworms (Read, 1951; Roberts, 1961; Ghazal and Avery, 1974). Recent work, summarized by Befus (1975), suggests that immune responses are involved in producing the crowding effect. However, the exacerbating effects of low carbohydrate diets and the ameliorating effects of high carbohydrate diets (Read, 1959; Roberts, 1966; Mead and Roberts, 1972) strongly suggest that the basic mechanism involved is exploitative competition, and that the role of immunity is secondary.

Intraspecific interference competition may be involved as a regulatory mechanism where infrapopulations are limited to a very small number of individuals. With the more complex parasites, such as mites (Mitchell, 1965) or hymenopterans (Salt, 1959), the aggression may be overt. In other cases, chemical interference seems more likely. For example, infections of *Gyrocotyle* in individual chimaerids are usually precisely limited to two worms (Halvorsen and Williams, 1968; Simmons and Laurie, 1972). Halvorsen and Williams (1968) have suggested that the limitation may be due to "conditioning" of the habitat by secretions from the helminths. Such a situation appears to be similar to the more well-known cases of chemical interference, such as allelopathy in plants (Rice, 1974).

Interspecific interactions can obviously affect the sizes of parasite populations, and may possibly be involved in regulating them. Predator-prey interactions are apparently uncommon, but may be important in those parasites (or endocommensals) which feed on other organisms in the digestive tract, such as some rumen-dwelling entodiniomorph ciliates, which feed on others (Hungate, 1975).

Interspecific exploitative competition can have the same effects as intraspecific competition (Holmes, 1961; 1962). Interference competition may also operate on an interspecific level, as suggested by the limitation of infections of *Gyrocotyle* to only one of the two species available (van der Land and Dienske, 1968; Simmons and Laurie, 1972). In either case, where two such competitors are common in a system, their infrapopulations may well be regulated conjointly. The ratio between them, and hence the suprapopulation of each, could be determined by their relative capacities for transmission and survival, in a manner similar to that suggested by Horn and MacArthur (1972) for competing species on islands. (Note that this would connote another kind of Type II regulation).

It should also be emphasized that regulation at the level of the infrapopulation, whether through immunological mechanisms, competition, or a combination of the two, is not necessarily accomplished through a limitation on the numbers of parasites in the infrapopulation. It may also be accomplished through a limitation on the reproductive output of that infrapopulation. Fecundity may be regulated through reductions in growth rates and biomass, as in various hymenolepidids (summarized in Ghazal and Avery, 1974) or through inhibition of maturation (Wisniewski, Szymanik, and Bazanska, 1958; Kethley, 1971; and references in Schad, this volume).

In most cases, the mechanism of negative feedback operates in a continuous fashion: the higher the number of parasites taken in, the lower the proportion which establish initially (Dobson, 1974), the higher the mortality (Burlingame and Chandler, 1943; Sutherst, et al., 1973), the lower the proportion that mature (Wisniewski, Szymanik, and Bazanska, 1958; Dunsmore, 1960; Michel, 1974), or the lower the growth rate (Roberts, 1961; Holmes, 1962) or reproduction per adult (Krupp, 1961; Kethley, 1971; Ghazal and Avery, 1974).

In some cases, however, regulation may be through a rather precise ceiling on the number of reproducing adults in an infrapopulation, as in *Gyrocotyle* (references above) and some other helminths (reviewed in Halvorsen and Williams, 1968; Williams and Halvorsen, 1971).

Therefore, we can recognize three distinct mechanisms of Type III regulation, plus two modes of action. Regulation may be through modification of
1) the number of parasites which establish, or which survive to reproduce,
2) the proportion of the surviving parasites which reproduce, or
3) the number of eggs produced per reproducing parasite.

The modification may act in a continuous fashion (all three mechanisms) or by establishing a ceiling (the first two mechanisms).

The regulation of a parasite's suprapopulation becomes even more complex when that population is spread over several species of alternative hosts. In many cases, the basic stock of the parasite is maintained in one host species (the primary host, or donor) from which it spreads to other host species (the secondary hosts, or recipients). *Fascioloides magna* (Bassi), in donor deer and recipient sheep or cattle (Swales, 1935), is a familiar example. In this case, the donor species is the only one in which the parasites reproduce to any significant extent, and is clearly the only definitive host involved in regulating the suprapopulation. Similar situations have been described by Hibler and Adcock (1971) and Anderson (1972).

In other situations, it may not be so easy to tell which is donor and which recipient. Where sufficient data are present, it is customary to follow Michajlow (1959) or Dogiel, Polyanski, and Kheisin (1964) and use the prevalence (percent infected) and/or intensity (average number of individuals per infected host) of infection, or the reproductive potential of individual parasites, as indicators of the primary hosts. However, the important criterion is the relative rate of flow of parasites through each host population, which involves measures, such as the relative sizes of the host populations and the turnover of parasites in them, not considered by those workers.

A good example is that of *Schistosoma japonicum* in the Phillipines, as summarized by Hairston (1962). Four species of mammals (humans, dogs, pigs and field rats) acted as definitive hosts. In humans, the prevalence of schistosomes was very high and female schistosomes had their greatest life span and total egg output. In dogs, the prevalence was only about half that in humans and female schistosomes had a shorter life span, thus only about 5% of the total egg output as in humans. However, the intensity of infection (as evidenced by the daily egg production per host individual) was three times that in humans. In pigs, the prevalence was about three-fourths and the intensity about one-fourth that in humans, and the life span very short, thus the egg output per female was less than one percent that in humans. In field rats, the prevalence was equal to that in humans, but the intensity only one-third, and the egg output per female less than 0.1% that in humans. By Michajlow's and Dogiel's criteria, either humans or dogs would be the primary host. However, based on an actuarial approach, Hairston (1962) suggested that the suprapopulation of *S. japonicum* in that study area could not be maintained by infections in humans or in dogs, but could apparently be maintained by infections in field rats, in which a high host population and a rat-snail contact "several orders of magnitude more intimate" than human-snail contact, lead to a much higher rate of flow of parasites. In this system, the field rats appear to be the true primary host.

Situations such as the above, in which the parasite's infrapopulation dynamics are quite different in different hosts, pose the obvious question of how significant the regulation in each of these hosts is to the overall regulation of the suprapopulation of the parasite. Pushed to its limit, the question becomes, "Can regulation in one host species regulate the suprapopulation?" The question is an important one. Some of the most important of the parasites of man or domestic animals (such as schistosomes) have alternative, or "reservoir", hosts. It is important to know in which host to attempt artificial control. In addition, different types of regulation may be involved, and Bradley (1972) has shown that different artificial control measures are appropriate to each type.

In the balance of this paper, we will investigate that question. We will summarize a field situation in which each of Bradley's types of population determination may be operating but in different hosts; propose the hypothesis that regulation of the reproductive output of infrapopulations in the primary host only is sufficient to regulate the suprapopulation of the parasite; develop a mathematical model to examine the hypothesis and the conditions under which it may apply; extend the hypothesis to cover other mechanisms of regulation of parasite populations; and discuss the applicability of the hypothesis to some current literature.

THE FIELD EXAMPLE

Cold Lake is a large (surface area, 373 km^2), deep (mean depth, 60 m; maximum depth, 100 m) oligotrophic lake located about 290 km NE of Edmonton, Alberta (Canada). It is ice-covered from mid-December through late May; water temperatures during this period are 1-2°C. During the summer, there is a well-defined thermocline, at a depth of 10-20 m, with surface temperatures reaching 15-18°C and hypolimnetic temperatures of 4-8°C (Paetz and Zelt, 1974). The fish fauna is dominated by salmonids (lake whitefish, *Coregonus clupeaformus* (Mitchell), and cisco, *Coregonus artedii* (LeSueur)) and sticklebacks (*Pungitius pungitius* (L.)).

Leong Tak Seng examined 3132 fish belonging to ten species from Cold Lake. The results of this study are available in Leong (1975), and will be published elsewhere; here we present a summary sufficient only to develop the hypothesis examined in the model. The acanthocephalan, *Metechinorhynchus salmonis* (Muller) was the dominant parasite, infecting all ten species of fishes (Table 1) and comprising 75% of all parasites

TABLE I.

Hosts of Metechinorhynchus salmonis *in Cold Lake, Alberta. (Data from Leong 1975).*

	Number Exam.	Percent Infected	Mean No. Worms	Percent Gravid Female Worms
SALMONIDAE				
LAKE WHITEFISH	863	99	170	14
CISCO	757	32	9	50
LAKE TROUT	35	100	413	24
COHO (AGE II)	210	99	58	20
COHO (AGE III)	78	100	533	57
ESOCIDAE				
NORTHERN PIKE	62	84	31	3
GADIDAE				
BURBOT	29	100	174	2
CATASTOMIDAE				
WHITE SUCKER	36	8	4	0
LONGNOSE SUCKER	12	83	31	0
PERCIDAE				
WALLEYE	12	67	3	0
GASTEROSTEIDAE				
NINESPINE STICKLEBACK	1038	19	2	0

recovered. However, it is apparent from the data in Table 1 that only the salmonids are important hosts for *M. salmonis*; the worms in non-salmonids showed little gonadal development, and the few gravid females are considered (for reasons presented elsewhere) to be worms which transferred from prey cisco to the predators (pike and burbot).

Whitefish are bottom feeders which feed extensively on the amphipod *Pontoporeia affinis* (Lindstrom), the intermediate host of *M. salmonis*. Three lines of evidence suggest a density-dependent regulation of the maturation of the acanthocephalans in individual whitefish (Type III regulation, operating on the reproductive output, not the infrapopulation itself). First, the mean number of gravid females per fish remained fairly constant (minimum 9.1, maximum 17.3, overall mean 11.9) in fish of age classes IV through IX despite a linear, almost fivefold increase in mean total numbers of acanthocephalans per fish (59 worms in fish of age class IV, 238 in fish of age class IX). Second, the mean number of gravid females per fish remained fairly constant between months (3.9 - 16.6, overall mean 10.8), with no obvious pattern throughout the year, despite a marked seasonal variation in total numbers of acanthocephalans per fish (monthly means of 83 - 348, with well-defined peaks in mid-summer and mid-winter). Third, and most instructive, there is a significant negative regression of the percentage of gravid females on total numbers of acanthocephalans in individual fish.

Cisco are primarily planktivorous, feeding only occasionally on *P. affinis*. Only a third of the cisco were infected, with low infrapopulations of *M. salmonis*, but there was a high percentage of gravid females (Table I). This appears to be an example of Type I determination of the acanthocephalan population, through transmission, with no negative feedback system in operation.

Lake trout (*Salvelinus namaycush* (Walbaum)) are predators, feeding almost exclusively on fishes. They presumably acquire *M. salmonis* through cisco or sticklebacks, which function as transport hosts. They were all infected, with large infrapopulations of *M. salmonis*, but the percentage of gravid females was low (Table I). The lake trout also harbored substantial numbers (mean: 33) of a large cestode, *Eubothrium salvelini* (Schrank). The data are scarce, but there was a negative regression of the percentage of gravid females on the number of cestodes and on the number of acanthocephalans (with the former more pronounced), suggesting a Type III regulation through the effects of inter- and intraspecific competition on maturation.

Coho salmon (*Oncorhynchus kisutch* (Walbaum)) were introduced into the lake by Alberta Fish and Wildlife Division in the early summers of 1970 through 1972. Young coho fed extensively on *P. affinis*, older coho on fishes, especially sticklebacks. They became infected with *M. salmonis* soon after being introduced, and the overwintered coho (age III) harbored large infrapopulations (Table I). The percentage of gravid female worms was low in the recently introduced age II coho, probably due to a short period of exposure and a lack of time for maturation, since the percentage of gravid females increased toward autumn, and was high in the overwintered coho. No negative feedback was apparent in the data, but the high infrapopulations of *M. salmonis*, its known pathology in Eurasian salmonids (Petrushevski and Shulman, 1961), and the low survival of coho from these introductions suggest that the most heavily infected coho may have died, a Type II regulation. Thus, the determination in coho may be of Type I or Type II.

It should be emphasized that no experimental studies have been conducted to verify the suggestions made above or to look for other types of control measures. However, in its four salmonid hosts, *M. salmonis* appears to be unregulated (determined by transmission) in cisco and possibly in coho, regulated at the level of the host individual in whitefish (presumably by intraspecific competition) and in trout (presumably by inter- and intraspecific competition), and possibly regulated at the level of the host population in coho. In addition, considerable numbers of acanthocephalans were found in hosts in which their reproduction was negligible, thus acting as a loss to the system. What significance does each of these have for the regulation of the suprapopulation of *M. salmonis* in Cold Lake?

An answer to this question requires information on the relative rates of flow through each of these hosts. Flow rates were not measured directly, but a first approximation was deduced from a static picture of the distribution of *M. salmonis* in the various species of fishes. The relative flow to each host was calculated as the abundance (prevalence times mean intensity) in that host species, multiplied by a weighting factor (derived from catch statistics from a comprehensive survey of the lake) to account for the relative size of the population of that host, then adjusted to a proportion of the total flow to all host species. (Note that these flow rates do not include ingested worms which fail to establish). Values for abundance in whitefish and cisco have been adjusted to account for the age structure of the host populations and for a seasonal sampling bias (see Leong, 1975, for details). The results are shown on the arrows leading from *P. affinis* in Figure 1. The relative output of eggs from each host species was taken as the abundance in that host species, multiplied by the proportion of gravid females in that host, multiplied by the weighting factor, then adjusted to a proportion of the total output. The results are shown on the wide arrows leading to *P. affinis* in Figure 1.

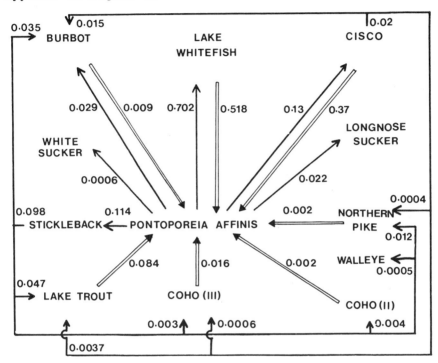

Figure 1. Derived relative flow rates for Metechinorhynchus salmonis *in a community of fishes in Cold Lake, Alberta. Outwardly-directed arrows indicate transfer with ingested* Pontoporeia affinis; *circumferential arrows indicate transfer with fish transport hosts; inwardly-directed (double) arrows indicate transfer of eggs to* P. affinis. *(See text for details.) From Leong (1975).*

The model can be simplified considerable by transferring the flow through transport hosts to the appropriate flow from *P. affinis*, combining the non-salmonids. The resulting model (Figure 2) suggests that most of the flow of acanthocephalans in the system is through whitefish. Since there appeared to be a well-developed Type III regulation of the maturation of the

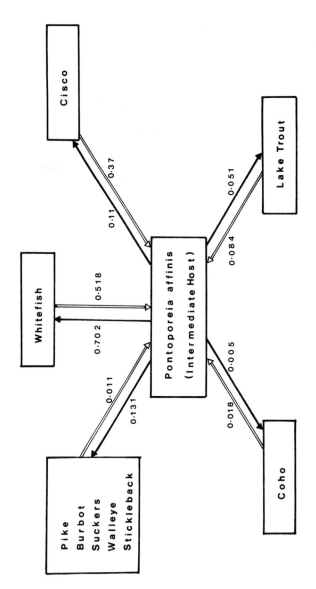

Figure 2. Simplified scheme of derived flow rates for Metechinorhynchus salmonis in a community of fishes.

acanthocephalans in that host, it further suggests the hypothesis that regulation of the major flow is sufficient to regulate the entire system. This hypothesis was examined, using a simple mathematical model.

THE MATHEMATICAL MODEL

The simplest mathematical model appropriate to the material, and the one which shows the relationships between parameters most clearly, is a deterministic model employing discrete generations. Each of the host groups in Figure 2 is represented by a compartment, with flow rates from the central compartment (the intermediate host) to the peripheral compartments (the definitive hosts) given by:

$$A_{ij} = I_j CF_i / H_i, \quad (1)$$

where
- A_{ij} = the number of acanthocephalans per fish of species i in generation j,
- I_j = number of acanthocephalan cystacanths in intermediate hosts in generation j, per fish (of all species),
- C = probability of a cystacanth being eaten by, and developing in, any fish (= total flow to fish),
- F_i = proportion of C going to species i, and
- H_i = proportion of fish belonging to species i.

The total of the flow rates from each of the peripheral compartments to the central compartment is given by:

$$I_{j+1} = \Sigma_i A_{ij} G_{ij} H_i E, \quad (2)$$

where
- G_{ij} = proportion of gravid female ancanthocephalans in species i in generation j, and

E = number of eggs per gravid female eaten by, and developing in, the intermediate hosts.

Negative linear feedback in the model is provided by having G_{ij} dependent upon A_{ij} according to:

$$G_{ij} = Q_i - R_i A_{ij}, \qquad (3)$$

where Q_i and R_i are constants for species i. This corresponds to the Type III regulation suggested for *M. salmonis* in whitefish and trout. (It should be apparent that the model can be used to investigate Type III regulation only: its deterministic nature precludes incorporating the Type II regulation suggested for *M. salmonis* in coho. Because of the behavior of the model, outlined below, this is not a critical weakness, at least for our immediate purposes).

The model assumes that the sizes of the host populations are independent of parasite populations. This is equivalent to stating that the parasites do not limit the populations of their hosts. In the system we are modelling (but not necessarily in other systems) we believe this to be a valid assumption (except, perhaps, for coho). For simplicity, we have considered the size of each host population to be constant. The model also assumes that the number of eggs produced per female, the probability of an egg infecting an intermediate host, and the probability of the resulting cystacanth becoming established in any definitive host species are independent of the species of host of the parent female. We have no information that suggests otherwise, so these also appear to be reasonable assumptions.

Another assumption of the model is that the probability of a cystacanth becoming established in a definitive host is independent of the number of cystacanths (or acanthocephalans) present. This is equivalent to stating that crowding does not affect the proportion of ingested cystacanths that become established, which appears to be the case for acanthocephalans in fish (Awachie, 1966; Kennedy, 1972; 1974). However, the assumption may not be correct for the very large numbers of acanthocephalans encountered in trout and coho.

The model also includes the simplifying assumption that interspecific interactions between parasites have a negligible effect on the establishment or maturation of *M. salmonis*. Although this assumption is invalid for infrapopulations in trout (as indicated above), the sacrifice of realism for clarity appears justifiable.

Some of the properties of this model (for stable point solutions) can be determined analytically. Substituting equation (1) into equation (2), and extracting the constants from behind the summation sign, we get:

$$I_{j+1} = CEI_j \Sigma_i G_{ij} F_i. \qquad (4)$$

The system can be regulated to a stable point by negative feedback operating on part of the system (e.g., on infrapopulations in whitefish) only if the number of cystacanths produced by eggs from the part of the system which is not regulated (i.e., infrapopulations in cisco, trout, and coho) is less than the number of cystacanths in the preceding generation. In mathematical terms,

$$CEI_j G_{max} F_N < I_j, \qquad (5)$$

where $F_N = F_i$ for all i in which G is unregulated. Assuming a 1:1 sex ratio, the maximum G if all females mature is 0.5. Substituting and solving for F_N,

$$F_N < 2/CE. \qquad (6)$$

Note that in this system, CE corresponds to the rate of increase of the acanthocephalans, so that equation 6 states that the flow of females through unregulated hosts ($F_N/2$), times the rate of increase must be less than one.

A second (and obvious) condition necessary for regulation (by any mechanism) is that the maximum number of cystacanths produced without any negative feedback must be larger than the number in the preceding generation:

$$CEI_j G_{max} (F_R + F_N) > I_j, \qquad (7)$$

where $F_R = \Sigma F_i$ for all i in which G is regulated (in our example, infrapopulations in whitefish). Solving for flow rates,

$$F_N + F_R > 2/CE. \qquad (8)$$

The equilibrium (stable point) maturation rate (G_R^*) can be predicted if it is regulated in only one species of host, as in our example, so that $\Sigma_i G_{ij} F_i = G_R^* F_R + G_{max} F_N$. At equilibrium, $I_{j+1} = I_j$. Substituting in equation (4), and solving for G_R^*,

$$G_R^* = \frac{1}{CEF_R} - \frac{G_{max} F_N}{F_R} : \qquad (9)$$

Equation (3) can be expressed in terms of G_R^*:

$$G_R^* = Q_R - R_R A_R^* , \qquad (10)$$

where A_R^* is the equilibrium acanthocephalan infrapopulation in the host where maturation rate is regulated (whitefish). Substituting (10) into equation (9), and solving for A_R^*,

$$A_R^* = \frac{Q_R}{R_R} + \frac{G_{max} F_N}{F_R R_R} - \frac{1}{CEF_R R_R} . \qquad (11)$$

The ratios between equilibrium acanthocephalan infrapopulations (A_m^* and A_n^*) in two host species (m and n) can also be predicted. Substituting equation (2) into equation (1),

$$A_{m,j+1} = \frac{CEF_m}{H_m} \Sigma_i H_i G_{ij} A_{ij} . \qquad (12)$$

At equilibrium values, j's disappear. Substituting X for the summation term,

$$A_m^* = \frac{CEF_m X}{H_m} , \text{ and} \qquad (13)$$

$$A_m^*/A_n^* = F_m H_n / F_n H_m . \qquad (14)$$

Therefore, given relative flow rates through the host populations (F_i), the proportions of the hosts in the system (H_i), the rate of increase of the parasite (CE), and the constants of the negative feedback equation (Q_i and R_i), it is possible to predict 1) whether control of maturation in infrapopulations in any one species could regulate the suprapopulation (by equations 6 and 8), and, if so, 2) the equilibrium maturation rate (by equation 9), and 3) the equilibrium infrapopulations in the definitive hosts (by equations 11 and 14). Given any four sets of values, such as

F_i, H_i, Q_i and R_i (which can be obtained from the field model), it is possible to predict the range of the fifth (CE) over which control is possible.

Predictions made in this way can be used to assess the effectiveness of regulation of the suprapopulation solely through control of maturation in a single host species. Two criteria seem appropriate: the range of conditions within which regulation can be effected, and the size of the equilibrium infrapopulations attained. The former is determined by only two variables (the rate of increase of the parasite and the rate of flow through hosts in which maturation is not regulated, and is shown by the shaded area in Figure 3. (The lower

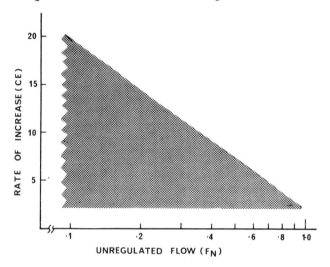

Figure 3. *The range of values of the rate of increase of the acanthocephalan (CE) within which regulation can be achieved (shaded area) at different rates of flow through hosts in which maturation is unregulated.*

limit in Figure 3 assumes that $F_R = 1 - F_N$; if hosts in which the parasite does not mature are included, as in the *M. salmonis* example, the lower limit would also slope upward to the left, with the slope dependent on the flow through such hosts.) It is apparent that this type of regulation is effective only when F_N is relatively low.

The sizes of equilibrium infrapopulations depend

on many more variables. Using the parameters from the field model, shown in Table II, the suprapopulation can be regulated by controlling maturation in infrapopulations in whitefish alone, but not by controlling maturation in those in any of the other hosts, alone or in concert. The maximum possible effect of adding control of maturation in one or more of the other hosts to that in whitefish would be equivalent to making maturation in that host zero. The results, using equilibrium numbers of acanthocephalans in whitefish (A_W^*) as a tracking variable, are shown in Figure 4. It is apparent

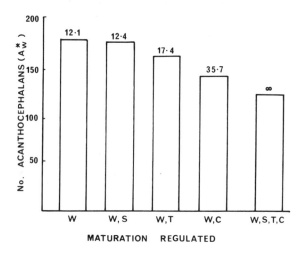

Figure 4. *Maximum effects on the equilibrium acanthocephalan populations in the primary host, whitefish (A_W^*), of regulation of maturation in the other hosts: coho (S), lake trout (T), and/or cisco (C). The maximum value of the rate of increase of the acanthocephalan (CE) within which regulation can be achieved is shown above each bar.*

that control of maturation in coho could have negligible effect (less than 1% reduction in A_W^*). (Therefore, the inability to model the possible effects of Type II regulation in this host, which would be less effective than zero maturation, is not a critical weakness of the model). Control of maturation in trout (a maximum 10% reduction) in A_W^* or cisco (maximum 10% reduction) could be somewhat more effective.

It would appear that the most important result of adding control in the other host species would not be the reduction in the sizes of the infrapopulations, but the greater range of rates of increase within which regulation could be effected. Upper limiting rates (CE values) are shown above each bar in Figure 4.

COMPUTER SIMULATION OF THE MODEL

In the simulations, we wish to check the conclusions reached analytically and to determine if there are any other conditions necessary for stability by comparing equilibrium infrapopulations of acanthocephalans under different sets of parameter values.

Simulations were run on an IBM 360 computer, using a program written in APL. The schematic flow chart for the program is shown in Figure 5. Standard values for parameters are shown in Table II. Values for H_i, F_i,

TABLE II.

Standard values for parameters and initial states used in the simulation models. See text for definitions of parameters.

Parameter	System	Value				
		Whitefish (W)	Cisco (C)	Lake Trout (T)	Coho (S)	Non-salmonids (O)
C	.1					
E	60					
H_i		.1477	.5182	.00083	.00021	.333
F_i		.702	.11	.051	.005	.132
A_i		32.33	1.45	413	186	2.48
Q_i		.5043	.5	.5	.5	0
R_i		.002143	0	0	0	0

and A_i were taken from the modified circulation model in Figure 2. Q_W and R_W are the constants of the line through $G_W = .5$, $A_W = 2$ (maximum maturation rate) and $G_W = .14$, $A_W = 170$ (means of the observed values for *M. salmonis* in whitefish). Maturation rates of the acanthocephalans in the other hosts were usually taken as maximum, although several runs were made with $Q_T = .501$ and $R_T = .0006205$, the constants of the line through $G_T = .5$, $A_T = 2$ (maximum maturation) and $G_T =$

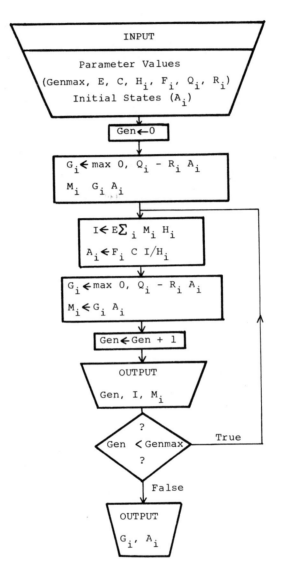

Figure 5. Schematic outline of the program for the model of population regulation through control of maturation. (Genmax = maximum number of generations in the run; Gen = current generation; G_i = proportion of gravid females; M_i = number of gravid females per host individual; I = number of cystacanths in intermediate hosts per definitive host (all fishes combined); other symbols as in Table 2).

.24, A_T = 413 (means of the observed values for *M. salmonis* in lake trout). (Similar lines for cisco and coho had negligible slopes). Components of the rate of increase (C and E) were set arbitrarily, but within the limits of stability predicted by equations 6 and 8, above.

The results of two runs with standard parameters- using the numbers of acanthocephalans per whitefish (A_W) and the proportion of gravid females in whitefish (G_W) as tracking variables, are shown in Figure 6. In

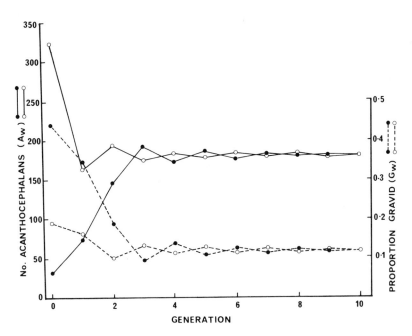

Figure 6. *The number of acanthocephalans per whitefish (A_W) (solid lines) and the proportion of gravid female acanthocephalans in whitefish (G_W) (dashed lines) in simulations with control of maturation on acanthocephalans in whitefish only. Parameters and initial values for Run 1 (solid symbols) as in Table 2; initial A_i values for Run 2 (open symbols) 10 times those in Run 1, other parameters as in Table 2.*

the second run, initial numbers of acanthocephalans (A_i) were ten times those in the first run. Each tracking variable showed rapid initial changes, followed by damped oscillations leading to a common equilibrium value.

The results of a large number of simulations, in which various parameters were altered, singly or in concert, substantiated the conclusions reached analytically and failed to reveal any other conditions necessary for stability. As expected from the analytic solutions, the simulations showed that equilibrium infrapopulations were not influenced by changes in initial populations (A_i), nor by changes in the components of increase (C and E), so long as their product (CE) remained constant. (Populations in the intermediate host were markedly affected, however). Equilibrium infrapopulations in the primary host, and their maturation rates, were independent of changes in the proportions of the host species (H_i), but infrapopulations in other definitive hosts were sensitive to such changes, as expected from equation 14.

Equilibrium infrapopulations in all hosts were sensitive to changes in flow rates (F_i) and the rate of increase (CE), as expected, but simulations revealed unexpected behavior of the model at high rates of increase. Stable-point equilibrium infrapopulations (solid curve) and maturation values (dashed curve) in the primary host, calculated from equations 11 and 10, respectively, are shown in Figure 7. Up through CE values of 6, the simulation results agreed with the predictions, but at values of 7 and above, infrapopulations appeared to reach limit cycles rather than stable points. The vertical bars at the higher CE values in Figure 7 show the magnitude of fluctuations in the limit cycles, as evidenced by maximum and minimum values from simulation runs. The oscillations appear to be due to the delay produced by the discrete generations, or, in effect, the integration interval (Maynard Smith, 1974). R. Hudson has pointed out to us that the smaller ranges of the oscillations at the highest rates of increase are due to the restriction of the maturation rate (G_w) to positive values (or zero). Fluctuations in maturation rate are thereby restricted more and more with increasing rates of increase.

It is also of interest to note that the maximum rate of increase which resulted in a stable limit cycle

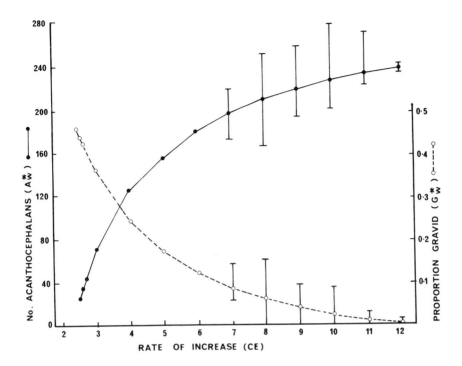

Figure 7. Equilibrium numbers of acanthocephalans per whitefish (A_W^*) (solid curve) and proportion of gravid female acanthocephalans in whitefish (G_W^*) (dashed curve) with different rates of increase (CE). The curves are for values calculated using parameters as in Table 2, using formulas 11 and 9, respectively. The vertical limes indicate boundaries of the limit cycles from simulations using the continuous feedback model.

in the simulations was the same as the maximum value predicted by equation 6, which assumes a stable-point equilibrium. Using the parameter values in Table I, the predicted maximum CE was 12.05. A simulation using CE = 12.0 resulted in a limit cycle, one using CE = 12.1 increased without limit.

EXTENDING THE HYPOTHESES

It is apparent that a Type III regulatory mechanism operating solely on infrapopulations in the primary host can regulate the suprapopulation of a parasite. To what extent does that conclusion apply outside of the system modeled?

Earlier, we recognized three mechanisms of Type III regulation: control of the number of parasites which establish or survive to reproduce, control of the proportion of survivors which reproduce, or control of the number of eggs per reproducing parasite. The model discussed above was worded in terms of the second mechanism, but the equations, and the conclusions, apply to the others with some modifications.

To model the effects of control of establishment or survival to reproductive age, three parameters need to be redefined:

C = probability of a cystacanth being eaten by any fish,

A_{ij} = the number of cystacanths ingested per fish of species i in generation j,

and G_{ij} = the proportion of cystacanths that establish or survive to reproduce.

Equations (6) and (8) take their general form:

$$F_N < 1/CEG_{max}, \text{ and} \qquad (6A)$$

$$F_N + F_R > 1/CEG_{max}. \qquad (8A)$$

Although the equations apply, their interpretations obviously change. Equations (1) and (11) through (14) now pertain to numbers of cystacanths ingested, not numbers of surviving acanthocephalans. The latter is now given by rather complicated product of equations (9) and (11), which in the previous model described the numbers of reproducing adults. Plotted against the rate of increase (CE), it takes the form of a humped curve; with other parameters as in Table II, it takes the form shown in Figure 8, curve B. Numbers of acanthocephalans in other hosts, however, are proportional to curve A (equation 14).

Three parameters also need to be redefined in order to model the effects of control on the number of eggs per reproducing parasite:

A_{ij} = number of reproducing adults per fish of species i in generation j,

COMMUNITY ORGANIZATION AND REGULATION OF PARASITES

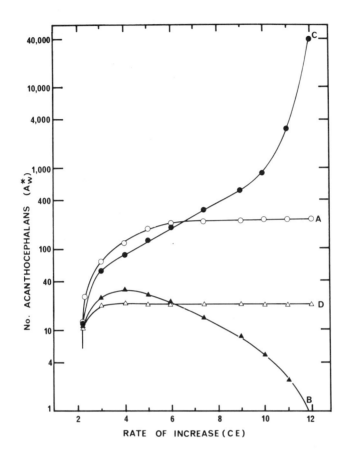

Figure 8. Equilibrium numbers of acanthocephalans per whitefish (A_W^*) as a function of rate of increase (CE), using different models. In all cases, regulation is through control on infrapopulations in whitefish only. Curve A: continuous feedback model, control of maturation or reproductive output per adult. Curve B: continuous feedback model, control of survival. Curve C: ceiling model, control of maturation. Curve D: ceiling model, control of survival.

G_{ij} = number of eggs produced per reproducing adult in fish of species i in generation j,

and E = probability of reaching and developing in the intermediate host.

Again, equations (6) and (8) take their general form. The rate of increase is now given by CEG_{max} and CE now gives the probability of an egg being eaten by and developing in an intermediate host, and being eaten by and developing in a definitive host. With these changes, the equations, and their interpretations, of the original model apply. Values of E, Q_i and R_W must be given new values to be meaningful. If this is done, keeping the value of CEG_{max} equivalent to that of the first two models and setting R_W so that $G_W = 0$ at the equivalent of $CE = 12.05$ (as in the previous models), A_W^* values from this model are identical to those of the original model.

The preceding analyses all assume that the regulatory mechanism acts in a continuous fashion. Earlier in this paper, we recognized an alternative mode of action, the establishment of a rather precise ceiling on the number of reproductive adults per infrapopulation. The relatively constant numbers of gravid female *M. salmonis* per whitefish throughout the year, and over a wide range of ages, suggested that this mode of action may be operating in our field example. The effect of this mode of action on the effectiveness of a regulatory mechanism (control of maturation) acting solely on infrapopulations in the primary host was tested in a second simulation model, shown schematically in Figure 9.

The analysis of the model is somewhat different, since equation (3) and its derivatives, equations (10) and (11), which are used to predict equilibrium infrapopulations in the primary host, do not apply. The rest do, so that the same conditions for stability exist, and the same equilibrium maturation rates (G_W^*) as in the original model will be attained with similar parameter values. However, the equilibrium maturation rates are attained by modifying the numbers of acanthocephalans present, not the numbers which mature, so that

$$A_W^* = \text{ceiling} / G_W^* \qquad (15)$$

Thus, the equilibrium numbers of acanthocephalans in the primary host (A_W^*), and those in other hosts, will be different from those of the original model.

Simulations using the parameter values in Table II,

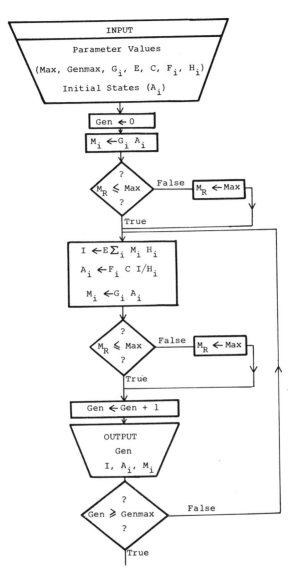

Figure 9. Schematic outline of the program for the model of population regulation through a ceiling on the number of gravid females. (max = maximum number of gravid females in whitefish, the regulated host; M_R = number of gravid females in whitefish; other symbols as in Figure 5).

and a ceiling of 20 gravid females per whitefish, gave the equilibrium infrapopulations (A^*_w) shown in Figure 8, curve C. It is apparent that regulation of the suprapopulation solely through a ceiling on the number of gravid females in infrapopulations in the primary host can be effective with low rates of increase, but not with high rates.

The "ceiling" model is also applicable to regulation by control of the number of surviving acanthocephalans. At all except very low rates of increase, the equilibrium infrapopulations in the primary host will be at the ceiling (curve D, figure 8). It should be emphasized, however, that infrapopulations in other hosts, in which the ceiling does not apply, and consequently the suprapopulation, will be proportional to curve C.

DISCUSSION

The most important conclusion of this investigation is that a suprapopulation of a parasite can be regulated through a Type III mechanism operating on infrapopulations in only one of several species of hosts. The host species involved does not have to be the one in which the parasite is most abundant, or even the one through which there is the greatest flow of parasites. All that is required is that the combined flow through the other host species be inadequate to maintain the suprapopulation. It is obvious, however, that a control operating on a relatively small flow would need to be very efficient in order to be effective; less efficiency would be required in a control operating on a relatively large flow. In addition, a control operating on a relatively large flow could be effective over a greater range of rates of increase. Thus, regulation through controls on infrapopulations in the primary host should be the most effective and the most likely to occur.

The required control may be provided by virtually any of the immunological or ecological mechanisms known to operate on the survival, maturation or reproductive output of the parasite. The precise mechanism does not appear to be crucial to the operation of the system, although mechanisms operating on the establishment or survival of the parasites may be somewhat less demanding on the ecosystem provided by the host than those operating on maturation or reproductive output.

The mode of action of the control mechanism appears to be more significant. Mechanisms acting through a ceiling on the number of surviving or reproducing adults in each infrapopulation lead to large suprapopulations at the high rates of increase. Corresponding infrapopulations, at least in other species of hosts, become unrealistically high; some form of control (a Type II selective mortality, if nothing else) would undoubtedly be invoked before populations reached those levels.

To what extent are our results dependent on the assumptions used in our models? The myriad assumptions used in devising the field circulation model have been used primarily to provide a set of reasonable values for the parameters used; the sensitivity of the population model to those parameters has already been assessed. Other assumptions, such as the constancy of C and E for the progeny of parasites from different host species, can be relaxed by incorporating additional variables in a more complex model. Preliminary analyses of such a model indicate that our basic conclusion is unaffected. The assumption of constant host populations has been relaxed in a model in which the flow per host individual (F_i/H_i) has been held constant, but host populations have been allowed to fluctuate by $\pm 10\%$. A preliminary analysis suggests that results can be drastically altered at high and low rates of increase, and are blurred at intermediate rates, but our basic conclusion was still unaffected.

Hirsch (this volume) has pointed out that two or more regulatory mechanisms may operate on the same species (or stage). In our model, only one mechanism of regulation was allowed to act at a time. Combinations of mechanisms (still operating on infrapopulations in the primary host only) could be expected to be more effective, or effective over a wider range of rates of increase, in regulating the suprapopulation, than a single mechanism. Alternatively, each mechanism could be less efficient than necessary to regulate the population by itself. An easily envisaged combination would be a control of the probability of a cystacanth becoming established dependent on the density of ingested cystacanths plus a control of maturation dependent on the density of established acanthocephalans. We have not attempted to model such a combination, but since the latter, added to our model of the former, would have the effect of lowering E, the combination would be expected to be more effective.

We have not attempted a stochastic version of the model. Such a version, particularly one which allowed a test of the hypothesis that a Type II selective mortality of heavily infected primary hosts could regulate the suprapopulation, obviously would be of interest.

An important feature of the mathematical model is that generations of the parasite are discrete with control measures acting independently on each. A model with overlapping generations, in which the effects of immune mechanisms might be more realistically modelled, also would be of obvious interest.

What are the consequences of this type of regulation of a parasite's suprapopulation for the hosts in which no regulation occurs? Let us examine an hypothetical situation in which equilibrium infrapopulations in the primary host were increased by an adaptation in the parasite which led to less density dependence or increased transmission rates.

In those host species for which the rate of flow of parasites per host individual (F_i/H_i) was low (i.e., cisco in our example), the increase in equilibrium infrapopulations would be relatively small. The added parasites may be more easily tolerated than defended against. The added evolutionary pressure to develop a mechanism to control the infrapopulations would be rather low, and one might expect that infrapopulations in such hosts would continue to be dependent on transmission.

In those host species for which the rate of flow of parasites per host individual is relatively high (i.e., coho and lake trout in our example), the increase in equilibrium infrapopulations could be relatively large. The added parasites may not be tolerated, and may provide strong evolutionary pressure to develop (or activate) a mechanism to control their numbers. Failure to do so may lead to death of the more heavily infected individuals and even to extinction of the host in that system.

We may have witnessed an equivalent situation in Cold Lake. The rate of flow of *M. salmonis* to the introduced coho was apparently relatively high, and there was apparently no mechanism in coho to control their infrapopulations. It seems reasonable to assume that the resulting high numbers of acanthocephalans played a part in the poor survival of the coho.

The literature (cited earlier) clearly shows that parasites may have different population dynamics in different species of hosts. We have argued that such differences can include differences in mechanisms of regulation. One consequence of such a conclusion is obvious, but frequently ignored. Laboratory investigations using domesticated species of hosts may not reveal regulatory mechanisms operating in nature. For example, the use of coho as a laboratory host would not reveal the negative feedback on maturation apparent in infrapopulations of *M. salmonis* in whitefish in Cold Lake.

The literature contains several examples of situations in which no mechanism of negative feedback can be shown to operate in particular host-parasite combinations. The best-studied example is *Pomphorhynchus laevis* (Muller) in fishes in the River Avon (Kennedy, 1972: 1974; this volume; Hine and Kennedy, 1974a; 1974b). However, most of the evidence against the presence of a mechanism of negative feedback is based on a study of dace (*Leuciscus leuciscus* (L.)), with critical evidence on the lack of density-dependent effects on establishment and survival based on experiments in goldfish (*Carassius auratus*) (Kennedy, 1972; 1974), which are not natural hosts of *P. laevis*. In addition, Hine and Kennedy (1974a) have pointed out that chub (*Leuciscus cephalus* (L.)), not dace, are the primary hosts for *P. laevis* in the River Avon, and that there is little apparent damage due to the presence of *P. laevis* in dace. Under such circumstances, the absence of a mechanism of negative feedback regulating infrapopulations in dace is not as surprising, nor as contradictory, as it first seems.

We do not know whether the flow rates for *P. laevis* in the River Avon are such as to make possible regulation of the suprapopulation through regulation of infrapopulations in chub, but we do suggest that such an hypothesis is worth investigating. (Since this paper was first presented, C. R. Kennedy (personal communication) has investigated *P. laevis* in chub and has found evidence of a density-dependent regulatory mechanism.)

We also suggest that those working with the population dynamics of parasites having alternative definitive hosts investigate relative flow rates through those hosts, and keep their populations in perspective when studying potential regulatory mechanisms.

ACKNOWLEDGEMENTS

This paper was written while the senior author was on sabbatical leave at Scripps Institution of Oceanography and the Department of Parasitology, University of Queensland. We thank J. F. Addicott, L. Cannon, P. K. Dayton, R. Hudson, D. Mauriello, R. Sutherst, and members of the parasitology discussion group at the University of Alberta for helpful comments. The study was supported by the National Research Council of Canada.

LITERATURE CITED

ANDERSON, R. C. 1972. The ecological relationships of meningeal worm and native cervids in North America. *J. Wildl. Dis. 8:* 304-310.

AWACHIE, J. B. E. 1966. The development and life history of *Echinorhynchus truttae* Schrank, 1788 (Acanthocephala). *J. Helminthol. 40:* 11-32.

BAILEY, N. T. J. 1957. The Mathematical Theory of Epidemics. Griffin, London.

BEFUS, A. D. 1975. Secondary infections of *Hymenolepis diminuta* in mice: effects of varying worm burdens in primary and secondary infections. *Parasitology 71:* 61-75.

BRADLEY, D. J. 1971. Inhibition of *Leishmania donovani* reproduction during chronic infections in mice. *Trans. Roy. Soc. Trop. Med. Hyg. 65:* 17-18.

BRADLEY, D. J. 1972. Regulation of parasite populations: A general theory of the epidemiology and control of parasitic infections. *Trans. Roy. Soc. Trop. Med. Hyg. 66:* 697-708.

BRADLEY, D. J. 1974. Stability in host-parasite systems. *In:* M. B. Usher and M. H. Williamson (eds.), Ecological Stability. Chapman and Hall, London.

BURLINGAME, P. L. and A. C. CHANDLER. 1941. Host-parasite relations of *Moniliformis dubius* (Acanthocephala) in albino rats, and the environmental nature of resistance to single and superimposed infections with this parasite. *Amer. J. Hyg. 33(D):* 1-21.

CROFTON, H. D. 1971a. A quantitative approach to parasitism. *Parasitology 62:* 179-193.

CROFTON, H. D. 1971b. A model of host-parasite relationships. *Parasitology 63:* 343-364.

D'ALESANDRO, P. A. 1962. In vitro studies of ablastin, the reproduction-inhibition antibody to *Trypanosoma lewisi*. *J. Protozool. 9:* 351-358.

DOBSON, C. 1974. Studies on the immunity of sheep to *Oesophagostomum columbianum:* effects of different and successive doses of larvae on worm burdens, worm growth and fecundity. *Parasitology 68:* 313-322.

DOGIEL, V. A., YU. I. POLYANSKI, and E. M. KHEISIN. 1964. General Parasitology. (English transl. by Z. Kabata). Oliver and Boyd, Edinburgh and London.

DUNSMORE, J. D. 1960. Retarded development of *Ostertagia* species in sheep. *Nature, Lond. 186:* 986-987.

ESCH, G. W., J. W. GIBBONS, and J. E. BOURQUE. 1975. An analysis of the relationship between stress and parasitism. *Amer. Midl. Nat. 93:* 339-353.

GHAZAL, A. M. and R. A. AVERY. 1974. Population dynamics of *Hymenolepis nana* in mice: fecundity and the "crowding effect". *Parasitology 69:* 403-415.

GRAY, A. R. 1965. Antigenic variation in clones of *Trypanosoma brucei*. *Ann. Trop. Med. Parasit. 59:* 27-36.

HAIRSTON, N. G. 1962. Population ecology and epidemiological problems. *In:* G. E. W. Wolstenholme and M. O'Connor (eds.), Bilharziasis. Ciba Foundation Symposium, Churchill, London.

HALVORSEN, O. and H. H. WILLIAMS. 1968. Studies on the helminth fauna of Norway. IX. *Gyrocotyle* (Platyhelminthes) in *Chimaera monstrosa* from Oslo Fjord, with emphasis on its mode of attachment and a regulation in the degree of infection. *Nytt. Mag. Zool.* 15: 130-142.

HIBLER, C. P. and J. L. ADCOCK. 1971. Elaeophorosis. *In:* J. W. Davis and R. C. Anderson (eds.), Parasitic Diseases of Wild Mammals. Iowa State Univ. Press, Ames.

HINE, P. M. and C. R. KENNEDY. 1974a. Observations on the distribution, specificity and pathogenicity of the acanthocephalan *Pomphorhynchus laevis* (Muller). *J. Fish Biol.* 6: 521-535.

HINE, P. M. and C. R. KENNEDY. 1974b. The population biology of the acanthocephalan *Pomphorhynchus laevis* (Muller) in the River Avon. *J. Fish Biol.* 6: 665-679.

HOLMES, J. C. 1961. Effects of concurrent infections on *Hymenolepis diminuta* (Cestoda) and *Moniliformis dubius* (Acanthocephala). I. General effects and comparison with crowding. *J. Parasitol.* 47: 209-216.

HOLMES, J. C. 1962. Effects of concurrent infections on *Hymenolepis diminuta* (Cestoda) and *Moniliformis dubius* (Acanthocephala). II. Growth. *J. Parasitol.* 48: 87-96.

HORN, H. S. and R. H. MACARTHUR. 1972. Competition among fugitive species in a harlequin environment. *Ecology* 53: 749-752.

HUNGATE, R. E. 1975. The rumen microbial ecosystem. *Ann. Rev. Ecol. Syst.* 6: 39-66.

KENNEDY, C. R. 1972. The effects of temperature and other factors upon the establishment and survival of *Pomphorhynchus laevis* (Acanthocephala) in goldfish, *Carassius auratus*. *Parasitology* 65: 283-294.

KENNEDY, C. R. 1974. The importance of parasite mortality in regulating the population size of the acanthocephalan *Pomphorhynchus laevis* in goldfish.

Parasitology 68: 93-101.
KETHLEY, J. 1971. Population regulation in quill mites (Acarina:Syringophilidae). *Ecology 52:* 1113-1118.
KRUPP, I. M. 1961. Effects of crowding and of superinfection on habitat selection and egg production in *Ancylostoma caninum*. *J. Parasitol. 47:* 957-961.
LEONG, R. T. S. 1975. Metazoan parasites of fishes of Cold Lake, Alberta: A community analysis. Ph.D. Thesis, Univ. Alberta, Edmonton.
MAYNARD SMITH, J. 1974. Models in Ecology. Cambridge Univ. Press, Cambridge.
MEAD, R. W. and L. S. ROBERTS. 1972. Intestinal digestion and absorption of starch in the intact rat: effects of cestode (*Hymenolepis diminuta*) infection. *Comp. Biochem. Physiol. 41A:* 749-760.
MICHAJLOW, W. 1959. Resistance of hosts and endoparasites, parasitological groups of hosts and types of systems "host-parasite". *Acta Parasitol. Polon. 7:* 467-487.
MICHEL, J. F. 1974. Arrested development of nematodes and some related phenomena. *Adv. Parasitol. 12:* 274-366.
MITCHELL, R. 1965. Population regulation of a water mite parasitic on unionid mussels. *J. Parasitol. 51:* 990-996.
PAETZ, M. J. and K. A. ZELT. 1974. Studies of northern Alberta lakes and their fish populations. *J. Fish. Res. Bd. Can. 31:* 1007-1020.
PENNYCUICK, L. 1971a. Quantitative effects of three species of parasites on a population of three-spined sticklebacks, *Gasterosteus aculeatus* L. *J. Zool., Lond. 165:* 143-162.
PENNYCUICK, L. 1971b. Frequency distribution of parasites in a population of three-spined sticklebacks, *Gasterosteus aculeatus* L., of different sex, age and size with particular reference to the negative binomial. *Parasitology 63:* 389-406.
PETRUSHEVSKI, and S. S. SHULMAN. 1961. The parasitic diseases of fishes in the natural waters of the USSR. *In:* V. A. Dogiel, G. K. Petrushevski, and Yu. I. Polyanski (eds.), Parasitology of Fishes (English transl. by Z. Kabata). Oliver and Boyd, Edinburgh and London.

READ, C. P. 1951. The "crowding effect" in tapeworm infections. *J. Parasitol. 37:* 174-178.

READ, C. P. 1959. The role of carbohydrates in the biology of cestodes. VIII. Some conclusions and hypotheses. *Exp. Parasitol. 8:* 365-382.

RICE, E. L. 1974. Allelopathy. Academic Press, N. Y.

ROBERTS, L. S. 1961. The influence of population density on patterns and physiology of growth in *Hymenolepis diminuta* (Cestoda:Cyclophyllidea) in the definitive host. *Exp. Parasitol. 11:* 332-371.

ROBERTS, L. S. 1966. Developmental physiology of cestodes. I. Host dietary carbohydrate and the "crowding effect" in *Hymenolepis diminuta*. *Exp. Parasitol. 18:* 305-310.

SALT, G. 1959. Experimental studies in insect parasitism. XI. The haemocytic reaction of a caterpillar under varied conditions. *Proc. Roy. Soc. (Biol.) 151:* 446-467.

SIMMONS, J. E. and J. S. LAURIE. 1972. A study of *Gyrocotyle* in the San Juan Archipelago, Puget Sound, U.S.A., with observations on the host, *Hydrolagus collei* (Lay and Bennett). *Internat. Parasitol. 2:* 59-77.

SMITHERS, S. R., R. J. TERRY, and D. J. HOCKLEY. 1969. Host antigens in schistosomiasis. *Proc. Roy. Soc. (Biol.) 171:* 183-194.

SUTHERST, R. W., K. B. W. UTECH, M. J. DALLWITZ, and J. D. KERR. 1973. Intra-specific competition of *Boophilus microplus* (Canestrini) on cattle. *J. appl. Ecol. 10:* 855-862.

SWALES, W. E. 1935. The life cycles of *Fascioloides magna* (Bassi, 1875), the large liver fluke of ruminants, in Canada, with observations on the bionomics of the larval stages and the inter-hosts, pathology of *Fascioloides magna*, and control measures. *Can. J. Res. 12:* 177-215.

TALIAFERRO, W. H. 1924. A reaction product in infections with *Trypanosoma lewisi* whcih inhibits the reproduction of the trypanosomes. *J. Exp. Med. 39:* 171-190.

VAN DER LAND, J. and H. DIENSKE. 1968. Two new species of *Gyrocotyle* (Monogenea) from chimaerids (Holocephali). *Zool. Med. Rijksmus. Nat. Hist. Leiden. 43:* 97-105.

WILLIAMS, H. H. and O. HALVORSEN. 1971. The incidence and degree of infection of *Gadus morhua* L.. 1758 with *Abothrium gadi* Beneden, 1871 (Cestoda: Pseudophyllidea). *Nor. J. Zool. 19:* 193-199.

WISNIEWSKI, W. L., K. SZYMANIK, and K. BAZANSKA. 1958. The formation of a structure in tapeworm populations. *Česk. Parasitol. 5:* 195-212.

INDEX

A

Abothrium gadi, infrapopulations and regulation of, 91
Abramis brama, serum antibodies in, 81
'Achievement factor,'
  and population density, 182
  and regulation of parasite populations, 74–76
Ancylostoma caninum,
  arrest, acquired immunity and environmental interactions, 145–148
  arrest and host immunological status (Table), 145
  arrest and host resistance, 124–126
  discovery of arrest in, 112
  immunity and arrest (Table), 147
Ancylostoma duodenale
  arrest and genetic variability, 131–132
  arrest and population dynamics, 153
Antibodies,
  in fish, 80–81
  precipitating, 5–6
Apodemus sylvaticus, and variations in parasite fauna in contrasting habitats, 18–19
Arrest (arrested development),
  definition, 111
  discussion of concept, 112–114
  discussion of physiological basis for, 148–150
  and dose-dependent effects, 137–141,
  environmentally induced, 114–123
  factors for induction of (Figure), 115
  and genetic variability, 130–132
  and geographic variability, 131–132
  immunosuppression and x-irradiation, 146–148
  and K-selection, 156
  as mechanism for population regulation, 112–114
  and natural selection, 153
  as option for ensuring survival, 113
  and regulation of helminth populations, 150–158
  and relationship to interacting variables, 141–148
  in relation to r-selection, 155
  seasonal variation in, 114–123
  and 'spring rise', 152–153
  suppression by adult worms, 132–137
  and temperature effects, 142–143
  temperature effects and immunity, 144–148
  theoretical considerations, 150–158

B

Behavior,
  alteration of host behavior by parasite, 21-23
  of cercaria, 21
  and parasite recruitment, 22-23
Bile salts, effect on *Echinococcus granulosus*, 14
Bunodera luciopercae, population dynamics and host age, 19
Bunodera sacculata, population dynamics and host age, 19

C

Camallanus oxycephalus, abundance and environmental fluctuations, 3
Capillaria caudinflata, survival time of, 46
Capillaria obsignata, survival time of, 46
Caryophyllaeus laticeps,
  and host feeding, 173
  model for regulation of, 185-190
  overdispersion of infrapopulations, 28
  regulation of infrapopulation (Figure), 185
  regulation of infrapopulation dynamics, 72-74
  and seasonal periodicity, 173
  and suprapopulation dynamics, 11
Chondrostomus nasus, parasite fauna in, 16
Chrysemys picta, parasites and diet, 16
Climate,
  K-selection and parasitism, 36-39
  r-selection and parasitism, 36-39
Colonization, by parasites, 43
Competition,
  allelopathy, 212
  and crowding effect, 4
  discussion of crowding effect, 87-91
  interspecific, 23, 86-87, 212-215
  interspecific and Type III regulation, 213
  intraspecific, 23, 87-91, 211
  intraspecific and Type III regulation, 217
  intraspecific for *Diplozoon paradoxum*, 71
  K-selection and parasitism, 44-45
  regulation and density-dependent natality, 89-91
  regulation by, 86-91
  r-selection and enteric nematodes, 32
  r-selection and parasitism, 44-45
Cooperia pectinata, arrest and host resistance, 128
Cooperia curticei, growth curve in sheep (Figure), 136
Crepidostomum cooperi, and infrapopulation overdispersion, 49

D

Dactylogyrus extensus, interspecific interaction with *D. vastator*, 87
Dactylogyrus vastator, interspecific interaction with *D. extensus*, 87
Dandelions, r-selection, 51
Density, of snails in relation to miracidium penetration, 191-194
Dictyocaulus viviparus, arrest and acquired immunity, 143-144
Diet, in relation to parasite recruitment, 16-20
Diphyllobothrium spp., as a cause of host mortality, 80
Diplostomulum scheuringi, metacercaria and temperature, 28
Diplostomum gasterostei, overdispersion and negative binomial distribution, 76-77
Diplostomum spp., as a cause of host mortality, 79
Diplozoon paradoxum, seasonal periodicity, 71
Dispersal, of miracidia, 190-194
Dispersion,
  and negative binomial model, 173
  and parasite recruitment, 28-29
Distribution, binomial probability, 192

E

Ecdysis,
  and developmental arrest, 148-150
  physiology of, 149-150
Echeneibothrium, establishment in definitive host, 14
Echinococcus granulosus,
  effect of deoxycholic acid on, 14
  establishment in crypts of Lieberkuhn, 14
Echinococcus multilocularis, establishment in crypts of Lieberkuhn, 14
Echinorhynchus clavula, overdispersion and negative binomial distribution, 76
Echinorhynchus truttae, and density-dependent regulation, 39
Egg production
  by *Macracanthoryhnchus hirudinaceus*, 1
  model for *Schistosoma japonicum*, 177

by *Moniliformis dubius*, 1
by *Polymorphus minutus*, 1
Epibdella (Benedenia) melleni, and mortality caused by, 82
Epistylis sp., season changes, 26
Epizootic, r-selection and parasitism, 51–52
Eubothrium salvelini,
  regulation and long-term stability, 92–94
  and Type III regulation, 217

F

Fasciola hepatica,
  lipid content of, 34
  and methods of control, 170
  model for regulation of, 190–194
  population biology and intermediate host (Figure), 191
  starvation and glycogen level, 34
  survival time of, 46
Fascioloides magna, survival time of, 46
Fasciolopsis buski, lipid content of, 34
Fecundity,
  and calorific levels of parasites, 33–36
  and evolution of endoparasitism, 33–36

G

Gambusia affinis, metacercaria infrapopulations in, 27
Gammarus lacustris, and parasite-induced behavioral changes, 22
Gasterosteus aculeatus, parasite population on dynamics in, 76–77
Gastrothylax cremenifer, lipid content of, 34
Genetic variability, and arrest, 130–132
Growth curve,
  for *Cooperia curticei* in sheep (Figure), 136
  for K-strategist (Figure), 31
  for r-strategist (Figure), 31
Gyrocotyle sp.,
  discussion of regulation for, 91
  and habitat conditioning, 212
  and interference competition, 212
  interspecific competition, 87
  regulation via chemical interference, 212
Gyrodactylus alexandrei, regulation of infrapopulations, 83–84

H

Haemonchus contortus
  arrest and genetic variability, 130
  arrest and host resistance, 128
  induction of arrest, 121–123
  population structure after experimental infection in sheep (Figure), 117
  regulation at suprapopulation level, 180
  seasonal changes in sheep (Figure), 117
Haemonchus contortus
  summary for regulation of (Figure), 180
  winter dormancy, 121
Haemonchus placei, suppression of development, 134
Heterakis gallinarium, survival time of, 46
Host acceptability,
  criteria for, 13
  and deoxycholic acid, 14
  inherent, 12–15
Host age,
  foraging habits and parasitism, 19
  and parasite recruitment, 18–20
Hyalella azteca, effects of parasites on development of, 2
Hymenolepis diminuta,
  density and stunting, 44
  and interspecific competition, 44
  lipid content of, 35
Hymenolepis citelli and Type II regulation, 4
Hymenolepis nana, and intraspecific competition, 44
Hypobiosis,
  definition of, 152
  discussion of, 152–153

I

Ichthyopthirius multifiliis,
  development and temperature, 26
  long-term host response to, 85
Immunity,
  acquired, and arrest, 144, 126–129
  against *Ichthyophthirius multifiliis*, 85
  arrest and age resistance, 123–126
  arrest and host sex, 126
  change with age, 19
  in chub, *L. leuciscus*, 6
  and competition in regulation, 212
  infrapopulation regulation, 80–82
  molecular mimicry, 15
  non-reciprocal, 15
  premunition and Type III regulation, 41
  regulation and parasitism, 50
  and regulation of parasite populations, 80–82
Infrapopulation,
  definition of, 11

discussion of, 63, 209

K

K-selection, 30-52
  and arrest, 156
  and climate, 36-39
  and competition, 44-45
  and longevity, 45-46
  and mortality, 39-41
  and population size, 43-44
  and survivorship, 41-42

L

Lacustrine ecosystem, succession in, 16
Leishmania donovani, and Type III regulation, 211
Lepeophtheirus pectoralis, population dynamics and temperature, 70
Lepomis macrochirus, and seasonal dietary changes, 17
Leptorhynchoides thecatus, seasonal changes in abundance, 18
Leuciscus cephalus,
  circulating antibodies in, 81
  precipitating antibodies in, 6
Leuciscus leusicus, and recruitment of *P. laevis*, 19
Ligula intestinalis,
  antibodies to, 81
  as a cause of host mortality, 79
  effect on host behavior, 78
  host-induced behavioral changes by, 22
Lipid content,
  of *Fasciola hepatica*, 35
  of *Fasciolopsis buski*, 35
  of *Gastrothylax cremenifer*, 35
  of *Hymenolepis diminuta*, 35
Longevity,
  Ascaridia dissimilis, 46
  Capillaria caudinflata, 46
  Capillaria obsignata, 46
  Fasciola hepatica, 46
  Fascioloides magna, 46
  Heterakis gallinarum, 46
  K-selection and parasitism, 46
  reduction by crowding effect, 45
  r-selection and parasitism, 45-46
  Syngamus trachea, 46
Longistriata adunca, and r-selection, 32

M

Macracanthorhynchus hirudinaceous, egg production by, 1
Metechinorhynchus salmonis,
  and evidence for density-dependent regulation and Type III regulation, 222
  infection percentages in Cold Lake (Table), 216
  populations, computer simulation for regulation of (Figure), 228
  regulation and consistency in numbers of gravid females, 234
  regulation by intraspecific competition, 88
  relative flow rates for (Figure), 219
  and Type III regulation, 6, 221-227
Micropterus dolomieui, plerocercoid migration in, 21
Model,
  computer simulation for regulation at suprapopulation level, 227-231
  for density-dependent regulation of *Taenia pisiformis*, 194-205
  descriptive type, 173-176
  deterministic, for regulation of *Caryophyllaeus laticeps*, 72-74
  deterministic, for regulation of parasite flow rates, 221-227
  deterministic, for regulation of parasite infrapopulations, 84-86
  deterministic versus stochastic, 175
  discussion of negative binomial, 171
  discussion of theoretical type, 172-176
  negative binomial, 171, 181-184
  negative binomial and parasite distributions, 48-50
  negative binomial distribution, 205
  predictive functions, 170
  program for model of population regulation (Figure), 224
  regulation of *C. laticeps* infrapopulation (Figure), 185
  for regulation of *Haemonchus contortus* populations, 180-182
  schematic for regulation at infra- and suprapopulation levels, 13
  second order polynomial, 199
  selection of parameters for, 172
  simulation and negative binomial, 173
  simulations for control of maturation (Figure), 228

simulations for equilibrium levels of
  acanthocephalans in whitefish (Figure),
  229
simulation values for parameters and initial
  states (Table), 227
steps for development of theoretical type,
  174
stochastic, for *Caryophyllaeus laticeps*, 72
stochastic, for regulation of parasite
  infrapopulations, 84
theoretical, for regulation of *Fasciola
  hepatica* populations, 190-194
types of, 172-176
Moniliformis clarki, and Type II regulation, 5
Moniliformis dubius, and interspecific
  competition, 44
Moniliformis moniliformis,
  egg production by, 1
  and Type II regulation, 5
Mortality,
  of hosts, induced by parasites, 79
  K-selection and parasitism, 39-41
  lethal level of parasite for host, 181-184
  as polynomial function of host density
    (Figure), 203
  rate for density-dependent related (Figure),
    198
  rate for density-independent related
    (Figure), 197
  r-selection and parasitism, 39-41

N

Negative binomial, for *Crepidostomum
  cooperi* metacercaria, 49
Negative feedback, discussion of, 236-240
Nematodirus spathiger, arrest and host
  resistance, 127
Nematodirus spp., seasonal changes in sheep
  (Figure), 117
Nematospiroides dubius, population
  dynamics in field mice, 18
Neoechinorhynchus rutili, and competitive
  exclusion, 45
Neoechinorhynchus spp., species diversity
  and temperature, 27

O

Obeliscoides cuniculi,
  arrest and infection dose, 138-141
  arrest and temperature, 139-141
  arrest versus larval dose (Table), 140
  induction of arrest, 121

Ornithodiplostomum ptychocheilus
  metacercaria and temperature, 27
Ostertagia ostertagi,
  adult worms and suppression of larval
    development, 132-137
  arrest and genetic variability, 130
  arrest and infection dose, 138
  environment and arrest, 114-123
  and 'spring rise', 152
  temperature and arrest (Figure), 120
Ostertagia spp.,
  arrest versus dose size (Table), 138
  seasonal changes in sheep (Figure), 117
Overdispersion,
  Crepidostomum cooperi metacercaria, 49
  and negative binomial model, 48
  negative binomial model and regulation
    of parasite populations, 72-74
  regulation and parasitism, 66
  and regulation of *Schistocephalus
    solidus*, 76-78
  r-selection and parasitism, 47-50

P

Parasite flow rate,
  discussion of, 218-221
  in fish communities (Figure), 219
  for *Metechinorhynches salmonis* (Figure),
    220
Perca flavescens, seasonal changes in parasite
  fauna, 19
Photoperiod, and arrest, 121-123
Plerocercoid, migration in relation to
  temperature and sexual maturity of
  host, 20
Polymorphus minutus,
  egg production by, 1
  frequency distribution and the truncated
    negative binomial, 171
Polymorphus paradoxus, induction of host
  behavioral changes by, 22
Pomphorhynchus bulbocolli, seasonal
  changes in abundance, 17
Pomphorhynchus laevis,
  antibodies to, 81
  and density-dependent regulation, 39
  and discussion of density-dependent
    regulation, 239
  and dynamic equilibrium in chub, 88-91
  recruitment by *Leuciscus leuciscus*, 19
  regulation by density-dependent natality,
    90

regulation by intraspecific competition, 90
and stimulation of precipitating antibodies, 6
Pontoporeia affinis, as an intermediate host for *Metechinorhynchus salmonis*, 216
Population dynamics, for population with one unstable and two stable equilibria (Figure), 205
Population size,
K-selection and parasitism, 43–44
r-selection and parasitism, 43–44
Posthodiplostomum cuticola, swimming behavior of cercaria, 21
Posthodiplostomum minimum, development in snail in relation to temperature, 25
Proteocephalus ambloplitis,
plerocercoid migration and temperature, 27
plerocercoid migration in bass, 21
and seasonal periodicity, 69
Proteocephalus filicollis,
and competitive exclusion, 45
and seasonal periodicity, 68
Proteocephalus fluviatilis, growth factors and temperature, 26
Proteocephalus stizostethi, seasonal periodicity, 17
Predation, of cercaria by intramolluscan larval stages of trematodes, 38

R

Recruitment,
active, definition of, 24
passive, definition of, 24
Regulation,
by density-dependent death and density-independent birth (Figure), 200
density-dependent, for *Taenia pisiformis*, 194–205
discussion of at the suprapopulation level, 213–215
and discussion of density-dependent growth, 169
discussion of mode of action of control mechanisms, 236–240
discussion of Type I, 40, 66
discussion of Type II, 40, 210–215
discussion of Type III, 41, 210–215
discussion of Type III at the infrapopulation level, 236–240
and extrinsic variables, 24–29
by habitat conditioning, 92
of helminth populations by arrest, 150–158

by host responses, 80–86
and intrinsic variables, 12–23
mechanisms of Type III, 211
model for *C. laticeps* infrapopulation (Figure), 185
model for *C. laticeps* suprapopulation, 185–190
model for control of *Fasciola hepatica*, 190–194
model for *H. contortus* suprapopulation, 180
model for *S. japonicum* suprapopulation, 176–179
by multiple density-dependent events, 202–203
and overdispersion of *Lepeophtheirus pectoralis*, 70
and overdispersion of parasite infrapopulations, 46–49
of parasite populations by serum factors, 81
at suprapopulation level, 11, 29
Type I, 4
Type I and regulation of parasite transmission, 217
Type I, of parasites in coho salmon, 218
Type II, 4–5
Type II, of parasites in coho salmon, 218
Type III, 4–6
Type III and reproductive output by parasites, 216
Reproductive strategies, r-selection and parasitism, 50
Reservoir host, concept of, 215
r-selection, 30–52
and arrest, 155
and climate, 36–39
and competition, 44–45
fecundity and caloric values of parasites, 33–35
and longevity, 45–46
and mortality, 39–41
and population size, 43–44
and survivorship, 41–42
Rutilus rutilus, behavioral alterations in due to parasites, 22

S

Schistocephalus solidus,
effects on predation of host, 23
overdispersion and log normal distribution, 77

Schistosoma haemotobium, infection and
host age, 19
Schistosoma japonicum,
dynamics of suprapopulation, 29
model for snail host portion of life cycle
(Figure), 177
regulation and suprapopulations, 214
regulation at suprapopulation level,
176-179
and suprapopulation dynamics, 11
Seasonal periodicity,
of *Caryophyllaeus laticeps*, 72-74
comparison of *Triaenophorus nodulosus*
populations from different localities, 69
of *Diplozoon paradoxum*, 71
for nematodes in sheep (Figure), 117
Selection, and arrest, 155
Sex,
relation to parasite establishment, 21
relation to parasite recruitment, 20
Stability,
concept of, 64-65
and regulation, 2
regulation of parasite infrapopulations,
92-94
Starvation, and glycogen depletion in
F. hepatica, 34
Stizostedion vitreum, seasonal changes in
parasite fauna, 17
Stress, r-selection and parasitism, 51-52
Succession, and parasitism, 17-18
Suprapopulation,
definition of, 11
discussion of regulation of, 209-215
and regulation, 29
Survivorship,
and cercaria of *T. patialensis*, 41-42
K-selection and parasitism, 41-42
r-selection and parasitism, 41-42
Type I curve, 42
Type II curve, 42

Type III curve, 42
Syngamus trachea, life span of, 46

T

Taenia pisiformis,
model for regulation of, 194-205
regulation of larval stages (Figure), 195
Temperature,
and arrest, 139
developmental arrest, 114-123
and development of larval trematodes, 25
and development time of protozoan
parasites, 26
effect on *Caryophyllaeus laticeps*
infrapopulations, 72-74
effect on migration of plerocercoids, 27
effect on tapeworm eggs, 26
effects on developmental arrest of
Ostertagia ostertagi (Figure), 120
role in regulation, 25-28
and species diversity of parasites in
turtles, 27
thermal effluent and parasites of bass, 27
thermal effluent and parasitism in
mosquitofish, 27
thermal effluent in aquatic ecosystems, 27
Transversotrema patialensis, and cercaria
mortality, 42
Triaenophorus nodulosus, as a cause of
host mortality, 78
Trypanosoma brucei, and Type III
regulation, 211
Trypanosoma lewisi, and Type III
regulation, 211

V

Variation,
phenotypic, 2
spatial, 2

THE LIBRARY
ST. MARY'S COLLEGE OF MARYLAND
ST. MARY'S CITY, MARYLAND 20686

083073